For Johnny Johnson—
With highest regard
and best wishes
always—
Andy

,1977

For The Common Defense

For The Common Defense

Andrew J. Goodpaster
General, U.S. Army (Retired)
The Citadel

Lexington Books
D.C. Heath and Company
Lexington, Massachusetts
Toronto

Library of Congress Cataloging in Publication Data

Goodpaster, Andrew Jackson, 1915-
    For the common defense.

    Includes index.
    1. United States—Armed Forces. 2. United States—Military policy.
3. United States—National security. I. Title.
UA23.G75   355.03'3073   77-4562
ISBN 0-669-01620-9

Published simultaneously in Canada.

Printed in the United States of America.

International Standard Book Number: 0-669-01620-9

Library of Congress Catalog Card Number: 77-4562

Contents

Preface

As Americans move into calmer seas after the tumult of the Vietnam years, we have an opportunity to review, reappraise, and reshape the contribution that is made by our military power to our national security. We have not only the opportunity, but also the obligation to do so. Changes in the world have occurred that have profound meanings for the future, and still further changes are in prospect. New questions are emerging, and clearer, more satisfying answers are needed for the problems of the past. The issues that face us range from the security goals and objectives that American military power should serve to the military weapons and the strength levels of military forces that will be best suited to serving these ends.

The evolving problems abroad—with adversaries, allies, and uncommitted alike—are inexorably bringing past solutions into question. Events and trends at home—which in recent years have set us at such severe cross-purposes with ourselves—have left much of the American public and its leaders unconvinced and uneasy about the essentiality, adequacy, and appropriateness of security policies and defense programs. Abroad, we need only to cite the uncertainty that exists regarding the purposes behind the continued Soviet military buildup and the massive Soviet military budgets, the internal difficulties that are to be seen in NATO and Western Europe (especially in the south along the Mediterranean flank), and the growing concern that is felt regarding the future course and prospects of the third-world nations of Asia, Africa, and Latin America. This concern includes worries about the dependability of future oil supply and the availability of other essential raw materials, as well as the hardening confrontation between the world's Northern and Southern Hemispheres, and the continuing slide of many third-world nations into nondemocratic systems of government.

At home, although the worst of the disunity and dispute from the Vietnam years is over, and although public support for a position of military strength "second to none" has reappeared, there are many questions and challenges as to just what that military strength and its specific weaponry should include. There is also considerable skepticism as to whether many of our ongoing military programs, commitments, expenditures, and new weapons systems are truly necessary or appropriate to the needs of our people and the present international environment. In the halls of Congress, as elsewhere, there is a frequent complaint that our military forces are not clearly and purposefully aligned to the policy objectives they should serve.

All this suggests that in the sizing and shaping of our military forces we should give added emphasis and more explicit attention to the *national-level purposes and security needs that these forces are to serve*. In relation to the pulls and pressures internal to the military establishment, the higher goals that should guide and shape defense programs more often appear to be the results of military programs than their determinants.

Of course this is not to deny the importance of such internal factors as the interplay of key personalities; individual and organizational ambitions and aims; the momentum of established programs; and the interests of affected groups ranging from defense contractors to congressional constituencies. For good or for ill, these factors—like the indisputable needs of our large military organizations for a considerable degree of continuity in personnel programs, in research and development, and in long lead time procurement and facilities usage—exert a profound influence on the dimensions, the content, and the timing of military programs and activities.

Nevertheless, in a review of military role and structure (specifically, the size and composition of our military forces and the concepts for their employment—i.e., what they should do, what they should be prepared to do, and how in broad terms they should do it), it is worthwhile to concentrate more heavily on external or purposive factors than on the internal factors. There are two main reasons: external factors often get less systematic and explicit attention than they deserve; internal considerations—requiring a body of current detailed knowledge—receive the attentions of the Defense Department, the Budget Bureau, and the Congress. In fact, the budget process itself fastens attention on specific programs and activities, rather than on a coherent, coordinated set of purposes that they should serve.

If power is indeed to be correlated with purpose, an approach that naturally commends itself is an analytic progression that moves from national values exposed to outside international threat to national interests requiring security protection, to security policies designed to safeguard such interests, and finally to military forces best suited by role and posture to carry out such policies, safeguard such interests, and protect such national values.

At this point, many would quickly object that the national values that give rise to security implications are so simple and self-evident as to need no specific mention. Others take the conflicting view that the values that guide and motivate us as a people are so many, so complex, and so amorphous as to defy codification—forming as they do the very substance of individual freedom and the political processes of self-government. But we may note that the *consequences* of such values, preferences, and aspirations—when they are stated in the more demanding terms of specific interests, policies, military implications, and possible attendant burdens and sacrifices—are neither so obvious nor so intangible. They are by no means free from uncertainty or from challenge, and this challenge can be resolved or reduced only by trying to assure that the policies and programs do in fact conform to and derive from the basic value judgments and priorities of our people.

A few words may be appropriate concerning the approach I have followed in trying to develop this line of argument. The examination is avowedly security centered and America centered. The reason is simply this: in the last analysis it is American security, however broadly it may be conceived, that provides the primary impetus and rationale for building and maintaining American military forces. The approach taken does *not* imply that nonsecurity (and nonmilitary) aspects of foreign policy are unimportant nor that the interests of other nations—particularly our allies—are inconsequential. To the contrary, it will be seen that consideration is given to them at many points in the analysis—although only, in general, as they bear on American interests and American security.

Americans have been undergoing an unusually difficult passage insofar as our security affairs are concerned, and there is little prospect that the problems will ease in the time ahead. While the problems of other countries are obviously worth careful study in their own right—and are of course receiving it today in many forums—the importance of the United States (and its military power) to world security is sufficiently great, and its role sufficiently difficult, to justify concentrating on it.

This book accordingly attempts to follow the sequence of inquiry just sketched out. It will suggest in rough outline how this approach appears to apply in the real world of today and the near future. Necessarily, the effort at application reflects to a certain degree my own judgments and observations, made over a number of years at a variety of governmental levels, and in a series of particular although diverse command responsibilities. Among the duties relevant to the purposes of this study were some 6½ years of service in the White House as a staff assistant on international matters to President Eisenhower; several tours of duty in the Joint Chiefs of Staff organization extending over a span of 20 years; command of a NATO-assigned U.S. mechanized division in Germany; 1-year service as deputy commander of our forces in Vietnam; and 5½ years as NATO's Supreme Allied Commander in Europe as well as Commander in Chief of U.S. forces there. Each one of these posts required a close look at one or more of the areas of inquiry involved in the present study. Each one also gave added emphasis to the importance of trying to correlate military power with national purpose. For the specifics of our military forces and weaponry, I have based my observations primarily on the Annual Reports of the U.S. Secretary of Defense and the Chairman of the Joint Chiefs of Staff to the U.S. Congress. In addition, I have referred extensively to the excellent study by the Library of Congress Congressional Research Service, *United States/Soviet Military Balance—A Frame of Reference for Congress* (U.S. Congress, Senate Committee on Armed Services, 1976).

Portions of the material contained in certain of these chapters have previously appeared in somewhat different form—respectively, as an article entitled "NATO Strategy and Requirements," *SURVIVAL*, International Institute for Strategic Studies (September/October 1975); as a Foreword, "Interests and Strategies in an Era of Detente: An Overview," *National Security and Detente* (New York: Thomas Y. Crowell Company, 1976); and as a section of a paper entitled "Educational Aspects of Civil-Military Relations," *Civil-Military Relations* (American Enterprise Institute Studies in Defense Policy). I am grateful to these publishers for their permission to utilize this material in the present study.

In addition, my gratitude is extremely great to the Woodrow Wilson International Center for Scholars, The Smithsonian Institution, Washington, D.C., for the opportunity afforded by my fellowship there in 1975 and 1976 to develop the thoughts contained in these pages. Particularly am I grateful to the staff of the Center for their unfailing support and to my colleagues for their comments and advice. I of course take full responsibility for the content of this study as it now appears.

Andrew J. Goodpaster
General, U.S. Army (Retired)
The Citadel, Charleston, S.C.
December 1976

For The Common Defense

1

American Security in Current Perspective

The elemental component of security for our country—assurance against harm and threat of harm—is known widely and grounded deeply. As for all times, it includes the safety of our people and of the territory in which they live; freedom from outside interference, coercion, and constraint in their internal affairs; an opportunity for the nation to survive and prosper in a world congenial to, or at least tolerant of, the purposes, aspirations, and activities of our people; the ability to pursue our values, and to maintain our institutions without hindrance by force from external sources; and a degree of confidence that all of these will be preserved into the indefinite future for the benefit of generations to come.

For America today, the needs of security take many increasingly complex forms, and these needs serve as the prime generator of both our military requirements and the wide range of American security policies and national efforts that are concerned with relations with foreign nations and institutions—friendly, unfriendly, or neutral in their posture and their attitudes toward us.

The foremost is, as always, the sheer physical protection of populace and homeland against hostility—a protection that peace provides but that war, no matter how skillfully conducted, could not hope to match, particularly if nuclear weapons were involved. War can, of course, be thrust on us; it takes two to make peace, or preserve it. And it can become necessary, if important values other than immediate physical safety should be threatened.

The concerns of security extend also to the safety of our commerce—especially the unimpeded access to sources of raw materials and other commodities on which our industry and the economic well-being of our people depend. The enormous flow of oil, food, and machinery across every ocean of the world—literally vital to the lives of the world's billions—depends critically on a world environment that provides a high degree of order, stability, and security.

And this country, like other countries, has an increasing security interest in other steadily multiplying linkages with the peoples of foreign lands—in exchanges of technological, scientific, cultural, and religious nature, and in open international travel for many purposes, which has become an extension of personal freedom into the international sphere.

The protection needed must be afforded against not only the possibility of direct military attack on the United States and its allies but also the adverse effects of military action elsewhere and the use of military threats by a hostile power to exert political pressure on us, whether directly or through other nations. The imposition of an alien control such as that of the USSR over one

1

after another of the weaker nations within the third world would cumulatively change the international balance to the severe detriment of our security position and security interests, and it would face us with security decisions of the utmost gravity. It would profoundly distort the international system as well, pressing it in the direction of major contending blocs, which sooner or later would be likely to come into armed confrontation.

National Values at Risk in the International Arena

Our basic security concern is the protection and preservation of American values—the safeguarding of our people in their persons, their associations, their possessions, activities, and aspirations—from outside harm. Our most enduring and compelling security interests are those that serve and safeguard, in essential ways, the deepest values and fundamental structures of our society. The richness and dynamism of our value structure doom to failure any attempt to codify it in comprehensive form. Nevertheless, the classic formulation of the national values which guide the federal government, set forth in the preamble to our Constitution two hundred years ago, forms a fitting foundation for our study:

... to form a more perfect Union, establish Justice, insure domestic Tranquillity, provide for the common Defence, promote the general Welfare, and secure the Blessings of Liberty to ourselves and our posterity. . . .

It identifies a concisely stated set of values—union, justice, tranquillity, defense, welfare, and liberty—that are hard to improve on for our purposes.

Broadly viewed in this context, American security becomes involved when and where such high national values are placed at risk in the international arena. The security equation thus includes elements both foreign and domestic: the American values that are subjected to risk; the foreign threats (or potential threats) that exist to such values; and the American responses to such threats—responses that help determine what residual risk will remain to the United States after practicable counteraction is taken.

We should note that the values to which our security efforts are dedicated can influence these efforts in several ways. First, such values constitute both the goals and the beneficiaries of such efforts. And while they are the objectives of such efforts, they can also serve as constraints on such efforts, in which case the latter must meet tests not only of effectiveness and of economy, but also of public acceptability and of public support as well. As an example, we, as Americans, would like to see the spread and the survival of democratic government in other countries of the world, and have in the past been prepared even to go to war for such purposes. The last few years, on the other hand, have seen a sharp decline in the nation's willingness to use force for such purposes either to protect or project democratic institutions, and certainly there exists today little thought of trying to impose such practices on hostile or unaligned nations. It is of interest, however, that we do attempt to press friends and allies in this direction—and that we will deny ourselves (and them) the benefits of cooperation in meeting common security needs if they seem to be violating such principles and norms. The thought frequently voiced is that such a sacrifice in immediate benefit will be more than balanced by adherence to the stronger, deeper foundation for such joint endeavor that greater insistence on democratic practices will help to achieve.

With regard to American values generally, it is fair to observe that with each passing year, the American activities subject to foreign interference seem to multiply and diversify at an accelerating rate. And the patterns of potential interference are themselves increasing. Whether our concern is at home, in foreign countries, or on, over, and under the high seas, the potential for harmful action is growing. It arises not only from foreign governments and their armed forces or other agents but, as we look to the future, from private organizations (such as terrorist groups) as well.

With respect to the United States itself, there remains the ever-present and overriding hazard of hostile military attack, and specifically of devastating nuclear attack, even though the present likelihood of such attack can—fortunately—be viewed as quite low. Beyond this the principal possibilities for foreign interference are to be found in economic initiatives such as the oil embargo, for which a military pattern of response by the United States has thus far attracted conspicuously little support. Externally supported subversion and terrorism remain further possibilities. Although they are not at this time of substantial proportions in relation to the total security threat, they may increase in the future.

Within the territories of foreign countries, the principal possibilities for security clashes involving the United States are to be found principally in those countries (notably South Korea and West Germany) where U.S. armed forces are deployed in direct proximity to military forces of the communist countries, or where U.S. interests and commitments would bring about U.S. actions in response to attack, threat, or interference from outside. In addition to Korea, our NATO Allies, Japan, and other friendly nations joined with us in security agreements provide prime examples. Within other foreign countries, there is a sharp limit to the possibilities that would have direct U.S. military implications, unless they involved the use of military force against U.S. activities and U.S. nationals. Expropriation and deprivation of liberty are matters typically dealt with in other ways—although even here it must be recognized that a possible chain of actions (refusal to give compensation, or seizure of ships and cargoes, for example) could lead to armed clashes.

On the high seas, the laws of piracy, backed up by strong, widely deployed Western naval power, have sufficed thus far to give a high order of protection to

American activities. The prospects for the future may be appreciably darker. Already, "cod wars" and shrimp-fishing controversies point to the possibility of violent clashes over economic exploitation of minerals, including oil, if a new law of the sea cannot be agreed on and implemented. In addition, clashes involving the United States could occur if unilateral attempts were made to close areas of the high seas, or international waters in the key straits of the world. As shown by Libya, sea mining can be a highly effective way to enforce such demands, and ambiguities between international law and national authority leave ample room to provide a potential threat of conflict. Increasing environmental pollution resulting from continuing industrial and population growth, whether emanating from the United States or affecting its activities in international waters, adds yet another source of future dispute. And the capabilities of terrorists to accomplish the destruction of aircraft in flight have been well demonstrated, offering a means for the weak to terrorize the strong that may prove irresistible if the spreading fears concerning frustrated, starvation-stricken countries of the world continue to materialize.

In the world's oceans, there is a kind of "no-man's land" in which the possibilities for dispute over deep-cruising submarines operating close to shore and the presence of listening or locating devices must also be recognized. For many years, tactics of harassment brought ships and aircraft close to dangerous accidents or deliberate damage. Happily, by the early 1970s, good sense had begun to impose restraint on such activities; agreement on rules of conduct was reflected in a welcome drop in the occurrence of such provocations and in a modest although significant improvement in the level and tone of the East-West relationship. Quick resumption of such tactics, should reversion to hard-line policies occur, remains of course a constant possibility.

Our country's expanding involvement and exposure abroad pose to the American people, and especially to each new administration as it comes to office, fundamental questions as to the level at which we should set our security interests. The military implications are substantial. Neither extreme holds much attraction. Although a middle ground is needed, it is not easy to find.

The high level would correspond to a strong outward involvement in world affairs. National consciousness of, and attention to, security issues would be high and alert. The "threat" posed from abroad—by the posture of possible military opponents or by the weakness and vulnerability of lesser nations—would be a matter of sharp and genuine concern. The anxiety over nuclear weapons—the desire to strengthen restraints on their possible use, and to forestall, delay, and inhibit the spread of nuclear weapons capabilities to additional countries—would be a prime motivator of national policy. The desire for strengthened stability, more effective world order, and the dissuasion of change other than by peaceful or internal processes would be reflected in readiness to commit national influence and resources to these ends.

We would recognize the interdependence of nations—a growing mutual dependence on the actions, the friendly attitudes, and the well-being of others for our own peace, security, and material progress. We would find economic, moral, and security satisfaction in seeing other nations free, prospering and nonaggressive in their outward policy. We would be prepared, in principle, to assist other nations to maintain their sovereignty and freedom from outside coercion, limiting ourselves in the actual burdens we could assume according to considerations of priority and feasibility. We would find a sense of reward and well-being in close, friendly association with other nations, especially the Western style industrial democracies.

We would be prepared to take positive action in support of the things that would strengthen peace and make it more durable, whether through deterrence or detente and particularly through reductions in the likelihood of nuclear war and the kinds of clashes that could lead to it. Where freedom was endangered, however, and a base of spirit and strength to defend it existed on the part of the people concerned, we would be prepared to risk military action to support such an effort, not only in areas of special significance to us—such as Western Europe and the Japan-South Korea complex—but in other areas of the world as well. In defense of freedom, "brush-fire" engagements and even the possibility of large-scale conflict (in Europe, for example)—preferably limited to conventional or conventional and tactical nuclear arms—would have to be recognized.

Our policy, for a time through the 1950s and well into the 1960s was widely thought to be as just described. The clarion call of President Kennedy, to "bear any burden" in behalf of the defense of freedom struck a strong chord of response. The intervention in South Vietnam was so perceived in its early years.

Following the actual test in Vietnam, however, together with the collapse of American policy and commitment there and the disenchantment of the American Congress over foreign burdens and risks, such a posture toward the outside world may be more demanding than anything Americans are likely to support. Because such a policy fails the test of feasibility and of credibility to our adversaries, our friends, and ourselves, it is necessary to search for more modest goals.

In sharp contrast to the high posture described would be a low posture, involving a minimum of the kinds of entanglement with the outside world that might have a military implication. Avoidance of nuclear war would remain the overriding aim, with reliance placed primarily on direct deterrence and detente arrangements, both efforts being supported by military power employable directly against the USSR (or Communist China, if necessary). Some commitments to friends—especially, the countries to which American ties of blood and identification are strongest—would be included as well. Beyond this, countries would be supported, if at all, by diplomacy, or left to fend for themselves with the support of world opinion, the United Nations, or other allies. An exception could be made in the states that are sources of oil, in particular, or of other important imports, if the continued availability of such supplies seemed to be threatened.

Such a formulation, limited as it is, nevertheless obviously goes well beyond the thinking of the U.S. isolationists of the 1930s (or their present-day successors) in envisaging continuing commitment to common security efforts such as NATO and the Japan-Korea arrangements, as well as the provision of strong forces to counter Russia and China. This policy probably lies within the bounds of feasibility, insofar as obtaining needed U.S. public support is concerned. This was demonstrated, after the Vietnam defeat, in the substantial military budgets that were approved by the Congress for 1976 and 1977, often over highly vocal opposition. Two questions must be raised, however. First, even when a policy of abandonment has been authoritatively decided on and announced, the actual event may occasion something quite different. Korea in 1950 and the Mayaguez incident in 1975 (quick commitments of U.S. armed forces in lieu of the inaction that might have been expected from earlier policy indicators) suggest what can happen, and they suggest as well the confusion and avoidable loss of American lives that can result from unprepared improvisation, however courageous and decisive the action may be when finally taken. But second, and more important, the chances appear high that such a policy in today's world could work only in the very short term. The outward thrust of the Soviets, and potentially of the Chinese, could well prove too strong—particularly if not counterbalanced by some sense of U.S. support—for the smaller nations of the world to be able to stand up to Communist-backed takeover regimes. These of course are not certainties. Perhaps the pulling power of nationalism and sovereignty, witnessed in Egypt and Syria in their assertion of national independence against Soviet presence and influence, will be enough. Nevertheless, this would be a policy involving considerable future risk.

In these circumstances, a recurrent thread that runs through American policymaking is the possibility of narrowing the American security interests that could involve our military forces, through deliberately curtailing our international activities or through strengthening the role of international institutions, international law, and norms of behavior. The difficulty here is that actual trends in the world today suggest the opposite to be far more likely. Today's national security interests and responsibilities derive from the growing aggregate of U.S. international interests, needs, activities, and objectives. These are not, in our society, just those dictated by the government. They arise directly and indirectly from the activities of many millions of our people and thousands of organizations, involved in international trade and travel, international mining and manufacturing activities, international finance and investment, international educational, cultural, and human welfare activity, and a myriad of other interactions.

There are inevitably certain hostages to fortune in all these interactions—at least, in all that create some dependency on independent and sovereign foreign decisionmakers for our safety and well-being. The U.S. writ does not run in the ways and places necessary to safeguard all these activities, and the UN is far from strong enough to provide a substitute—even if it were so inclined. Beyond this, the incidence of active or latent hostile intent toward the United States-particularly on the part of communists committed to making theirs the "wave of the future"—is sufficiently broad to provide an ever-present threat. These activities, and the dependencies and vulnerabilities they create, may be exploited to our detriment if opportunity and advantage for the communists present themselves.

The overall trend is thus a rising tide of economic, political, and intellectual cross-contact and a vast and growing number of mutual dependencies gathered under the title "interdependence." With this growth comes increasing vulnerability and potential loss, not only to deliberately harmful foreign action or violence but also to the mere denial of cooperation; the significance of the latter derives from the gains from international exchange, which have now become so great and so deeply embedded in national living standards and social progress. The risks are intensified by the growing sensitivity to disruption of modern economic and social life itself—the explosion of an aircraft in the sky, the burning of a communication switchboard center, the dynamiting of penstocks and power-line pylons, the sabotaging of tank-farms utilized for oil storage.

The extent to which international institutions and sanctions will be able to moderate, stabilize, and safeguard the burgeoning interdependent activities in the presence of these intensifying vulnerabilities is far from clear, but the pressures tending toward disintegration and destabilization are many and substantial.

The Widening Span of Security Problems

Since World War II and the early years of the Cold War, security problems have increased greatly in scope and diversity. The forms of possible conflict have multiplied. Characteristically, new aspects and issues of security have emerged, but the old, even if contained or reduced, have not wholly disappeared. Thus, where one of the greatest of our concerns in the 1950s was the possibility of overt Soviet military attack in Western Europe, or the seizure of control by Moscow-dominated regimes in the pattern of those in Eastern Europe, by the late 1950s and early 1960s, nuclear issues had come to the fore. The possibility of Soviet nuclear attack against the United States had expanded step by step with the rapid growth of their intercontinental and intermediate-range nuclear missiles and had brought a moment of nuclear confrontation in the midst of the Cuban missile crisis in 1962. Since then, emphasis has been placed on deterrence and the stability of the nuclear balance. As a result, many of our major concerns have tended to shift away from imminent nuclear attack, now rated much less likely, to a longer-term Soviet use of its military forces for purposes of threat or coercion, or for the progressive emplacement of Soviet military power (or Soviet-aligned regimes backed by Soviet military power) in the third world.

Since the time of Khrushchev, the Soviets have identified "just wars of national liberation" (widely regarded as no more than a euphemism for subversive overthrow of free governments) as a species of conflict to which Soviet assistance and possible Soviet participation might be provided. The United States, for its part, identified conflicts such as those in Korea and Vietnam as wars in defense of freedom, to which U.S. assistance and U.S. participation might be (and in those cases were) provided. Nor have military conflicts of other kinds been lacking. Local conflicts between neighbors were exemplified by Arab-Israeli and Indo-Pakistani warfare. Intermittent struggle of her African colonies against Portugal continued up to final Portuguese withdrawal. And the civil wars that continued to occur—in the Congo, in Nigeria, subsequently in Angola—have often been fueled with foreign support and have often become the focus of foreign involvement and intervention. Conflicts such as these seem likely to remain a real possibility on the world scene, with, on occasion, serious implications for U.S. national security.

But security involvement has spread more widely than such military clashes—or potential clashes. New kinds of challenges to our security, many of them not directly military in nature, have now arisen that have serious implications for the future. The vulnerability of the people of the United States—and of the whole Western industrial world—in their jobs and in their living standards to interference with the supply and availability of oil from the Middle East is only the most recent of the major nonmilitary threats that have been disclosed to us. The rising problems of population pressure, shortages of food, and unrelenting poverty in many quarters of the globe warn us of the possibility of widespread discord and dispute in the future between the rich and the poor nations of the world—clashes that may well involve by the 1980s international violence in many forms, together with increasing terrorism, of which the kidnappings of executives and political leaders and the seizure of industrial enterprises may be only early manifestations.

But it must not be forgotten that, along with the new, many of the old dangers still persist and have simply taken on new forms or new dimensions. The nuclear threat is of greater overall magnitude than ever, with the introduction of multiple warheads (MIRVs) and new types of cruise missiles, and it is aggravated by the proliferation of nuclear capabilities, most recently to India. The Soviet forces that overhang Western Europe continue to pose not only a military threat, but also an increasing power for political domination via the now-familiar processes of "Finlandization," forcing additional countries to submit to Soviet political pressures under the shadow of Soviet military force, just as Finland has had to do. And this at a time when NATO and the other security arrangements that have carried us well through the quarter century since World War II are showing signs of fatigue and in some cases of inability to come effectively to grips with the new security issues that are emerging, especially those outside the NATO treaty area. (Nevertheless, the mutual assistance provided by allies toward fulfillment of common security objectives, even if impaired in some respects, still remains a substantial contribution to the stability of vitally important areas of the world, and joint participation remains necessary and valuable to any adequate pattern of world order that we can realistically hope to see.)

To deal with the wider range of security problems that now confronts us, involving as they do far more than military threats, requires more than military measures of response and protection. A new and greater requirement exists to parallel our military capabilities and activities with processes and instrumentalities of a nonmilitary nature—political, economic, and other—and to place them all in a higher context of national purpose and policy.

A heavier burden of proof than in the past will rest on anyone proposing that U.S. military power be committed to deal with, or even to deter, a nonmilitary act damaging to American well-being. The Arab oil embargo—an act of economic warfare against the West—has shown in actual experience just how high the level of provocation may now be set without evoking military counteraction. Just how much higher the threshold is at which such counteraction *would* be taken is not known. Perhaps the seizure of control over the great oil reservoirs of the Middle East (in Saudi Arabia and Kuwait, for example) by regimes dominated by the Soviets and hostile to the West would suffice. But even here, if the future stance to be taken by these regimes were left ambiguous (as ambiguous, for example, as the future of Castro's relations with the United States was kept at the time of his takeover of Cuba), one may seriously doubt whether the nations of the West would find it within themselves to initiate military action.

Nevertheless, while military power is not the answer to all security needs and problems, there are many to which it remains quite indispensable. The forces and circumstances that have pulled the United States into closer, often more dangerous contact with foreign countries, and into deepening involvement, even entanglement, in international issues have taught us in two large wars and numerous smaller conflicts that the national interests and values involved are large enough to bring us into military participation. Moreover, we have seen—especially since World War II—that a particular pattern and design of armed strength, such as that in NATO, can reinforce stability and maintain security (as in Western Europe) without the outbreak of actual open conflict. American activities, interests, and needs abroad—on the high seas and in other countries—are important for political, economic, and strategic reasons to the continued safety, confidence, and well-being of our people. They are also important enough to the functioning of our economy, particularly our powerful industrial machine on which the standard of living of our people (and of much of the world) so vitally depends as to require military protection in case of threat. In an uncertain world of contending states with often conflicting ambitions, respect for our power—particularly our military power, with its ability to deny, to punish, or to compel—is an essential underpinning for our present and future welfare. It is likely to remain so for as long as our activities and interests continue under the potential challenge of other nations, the USSR foremost among them.

The responsibilities that fall to the United States cannot be narrowed simply to those that relate directly and immediately to the activities and interests of the United States alone. If the United States is to enjoy the benefits of peaceful finance and trade—in essential raw materials as well as remunerative exports— together with cooperative exchanges of a scientific and cultural nature, and the other benefits of reasonable world order, stability, and friendly association, it must play a major part in establishing and supporting a constructive world environment.

The Continuing Role of U.S. Military Power

For all of this, U.S. military power—together with a widely shared awareness of that power—remains essential. It is a major determinant of U.S. and world ability to sustain peace and security, with stability and national freedom. It helps to deter or constrain international resort to violence and to resolve quarrels and disputes when they occur.

With the need for military power and with the main components or categories of the required U.S. military power there can be little quarrel. Experience leaves us in no doubt that a degree of power balance, or military equilibrium, is essential if dangerous ambitions and expansionist urges are not to grow. In particular, it is necessary to maintain a strategic balance with the Soviet Union and to modernize and to strengthen our forces accordingly, while concurrently working—through arms limitation negotiations—to regulate and restrain the buildup on both sides. It is necessary as well to maintain a stable equilibrium in key areas such as the North Atlantic and Western Europe or the Western Pacific and Northeast Asia, working with allies in both regions. A further goal is to safeguard continued access to the Middle East. Sea lines of communication must be protected, along with major interests elsewhere if they should be endangered.

For these responsibilities and tasks, strong and modern forces—nuclear and conventional—are indispensable. They must be strong enough to deter armed challenge and to defeat it if it should come. They must be flexible enough to meet an ever-changing flow of specific problems, shifting alignments, and threatening disputes.

A crucial task in shaping the role and structure of American military power is determining how it should respond, or be able to respond, to the whole range of possible and foreseeable security needs. Ultimately, the size, composition, deployment, readiness posture, and planned strategic employment of such forces must be determined in the light of such considerations. For this purpose, the role and structure of these forces must be set within a well-defined framework of security interests and the policies to pursue them.

Formulating National Security Interests

Formulating national security interests forms one of the heaviest tasks of the national government, particularly at a time—as in the aftermath of Vietnam—when a new security balance must be struck and when domestic opinion remains unsettled and skeptical. A new formulation of national security interests must be developed and reconciled with other—sometimes competing, sometimes support-ing—segments of national interest and purpose. Important issues likely to affect the security of our people must be weighed in specific terms. Hard choices involving costs and trade-offs of many kinds must be made, ranging from better relations with one set of nations at the price of worse relations with another to improvements in one segment of our defense capability at the cost of cutbacks or loss of flexibility elsewhere.

While the national government, through its elaborate planning and decision-making machinery, assesses explicitly or implicitly security interests to be pursued, its determinations are not sacrosanct nor can they be static. Rather, they are subjected to intense scrutiny and challenge by individuals, interest-groups, bureaucracies, and molders of opinion. Often the representations of such groups are focused on single issues and the demands that particular participants are struggling to achieve. In the government itself, policy-through-accretion—the piecemeal accumulation of precedents through responses to short-term crises—often takes the place of more comprehensive weighing of priorities, of zero-base review, and realignment of action to broader purpose.

The task of accomplishing such a formulation of interests is admittedly complex and difficult. Nevertheless, it is a task that must be undertaken, on a basis widely shared. If security policy is to succeed, every major institution of our society—government, centers of learning, molders of public opinion, and the electorate itself from which ultimate authority flows—has the obligation in a world of strife and danger to form a workable and informed view as its contribution to programs and goals.

The flood of events and activities abroad that so affect and challenge us must be seen in some pattern if they are not to be a meaningless and frustrating jumble. They must be related to broader doctrine and concepts that enable us to deal with problems in larger frameworks. Strong pulls exist today that draw our people to a rapid-fire succession of emotion-gripping crises, tensions, and distresses. Without the help of orienting principles and purposes to add coherence and perspective, the possibilities of confusion, contradiction, weather-vaning, self-impairment, and failure are likely to become actualities.

Changing Challenges to National Security—At Home and Abroad

The challenges that present themselves to our basic security interests and needs are many, diverse, and subject to constant change. They are challenges that arise both from without and from within our own society.

Abroad, although much has been changing, much has nevertheless endured, albeit sometimes with major modifications. With the Soviet Union, the new relationship that is building has faced us with new problems of balance and perspective, many of them puzzling and unfamiliar, whose full security implications have not been fully disclosed. The wider, more open channels of communication between the Soviets and ourselves—reflected in an easing of nuclear confrontation and in results of profound historic promise in SALT—have been counterbalanced by a heavy continued buildup of Soviet military forces and by continuing Soviet efforts at penetration and expansion of their presence, influence, and hegemony in areas such as Africa and the Middle East—areas where important U.S. interests are involved and are adversely affected.

The "opening to China," coupled with the U.S. debacle over South Vietnam, has introduced far-reaching changes in U.S. security relationships in the Western Pacific and Asia as far west as India and Pakistan. These changes make it impossible to predict where and whether an enduring new equilibrium can be established. In Japan and South Korea the efforts to restore good relations following the shock occasioned by the U.S. approach to China attained a considerable success, but much will depend in the future upon the policies and ambitions of the successors to Chou En-lai and Mao Tse-tung as that troubled transition proceeds.

The new series of security problems that has been opening up—largely based on economic interests and differences—involves issues for which the traditional military strength of the major powers has little immediate relevance. In increasing numbers of third-world countries, policies of sharp antagonism toward the United States and other Western democracies have in recent years been displacing earlier friendlier relationships. And there is clear prospect that the intense population pressures, food shortages, and widening gaps between the people of the wealthy and the poor areas of the world that can now be foreseen will bring new gravity to a range of security issues that could threaten the destruction of any very stable world order.

The deepening dependence of the West—not excluding the United States—on Mid-Eastern oil, even after the producing nations convincingly demonstrated their ability and readiness to employ this resource as a powerful political weapon, has greatly altered security relationships, probably irreversibly. What is not yet known is whether other producers—of bauxite, for example—will be able to follow suit.

The traditional major alliances—NATO, in particular—retain their important security functions, but they have shown themselves ineffective in responding to challenges outside the old areas as defined by the treaties. What has been most lacking has been common purpose and the commitment to join in developing common lines of action—the very attributes that gave a strength to the alliance in its formative years far beyond the strength of its separate members.

The imminent spread of large quantities of nuclear fuel—inevitably resulting in the production of significant amounts of plutonium in many nations around the world—will, when coupled with the anticipated provision of uranium-enrichment and plutonium-separation technology as well as plutonium-cycle power reactors, increase the potential for nuclear proliferation and throw a darkening shadow over the existing nuclear balance.

The flagrant security failures and erosive trends that have occurred have not only damaged the security arrangements in major areas of Asia and the Pacific, but also they have lowered Western security confidence while enlarging the opportunities open to the Soviets. The final collapse in Vietnam showed clearly the wide gap between the earlier American commitment and the American effort available to fulfill it. The erosion of will and effort noted in the countries of Western Europe—with few exceptions—has been so widely remarked as to be undeniable. The drift of nations formerly dedicated to the building of representative governments into authoritarian, frequently military regimes steadily tilts the scales against the Western democracies—and removes an important bulwark against the imposition of Soviet-supported communist rule. The possibilities of national dismemberment and internecine conflict in Africa, held in check following the achievement of independence by doctrines supporting sovereignty and respect for existing borders, have been on the rise and were accelerated by interventions in Angola and the Spanish Sahara.

At home, events have necessitated a new appraisal of the domestic strength available to meet security needs. The experience of the Vietnam years makes clear that serious hazards and handicaps to our future security have emerged. Indeed, the challenges to our present and future security that come from within our society seem likely to be more weighty in the years ahead than those from without. In part, this situation has been the result of a shift in the national focus, which was timed with the bitter trauma and miserable result of the Vietnam War, with the convulsion of Watergate, with the breakdown of national consensus and lapse into governmental disarray, with the disappointment over the results of vast social programs despite immense governmental expenditures and staggering budgetary deficits, and with the persistence of economic disturbances—inflation, unemployment, and oil shortage—with which the normal processes of government have shown themselves to be embarrassingly ineffectual.

Insofar as foreign affairs, and particularly security affairs, are concerned, there has been noticeable corresponding turn away from a sense of participation and responsibility. The process was initially seen in a marked antipathy toward many military programs and a decline in the support accorded them. No one can of course say just how long-lasting this particular pattern of attitudes will prove to be. Some carry-over of less supportive, more inward-looking attitudes in relation to security affairs and foreign undertakings must surely be expected to persist for some time, however. A less engaged, more unilateralist tendency seems probable, and with this, by virtue of the memory of American collapse over Vietnam, there will be a residual lack of confidence as to whether the United States can or will show a high degree of endurance in meeting taxing security issues, unless U.S. survival itself is directly and unmistakably challenged.

The degree of public division, disorientation, and disbelief that has been evidenced regarding serious issues of security suggests that national leadership, which lost so much public trust and confidence in the late 1960s and early 1970s, will have a crucial role to play. The lack of unified leadership to chart a clear course or mobilize public support undoubtedly bears major responsibilities for the condition of internal disunity and disenchantment that has been so evident. One may hope this will prove to be only a passing phase, but if it is to be, remedial trends of a fundamental nature will be required.

For in many important circles—the Congress, the academic community, the news media, and among our youth—a continuing mood of disinterest, even suspicion and hostility toward the burdens and the involvements of security gives cause for concern. Even if more latent and less visible than in the Vietnam years, the basic attitude remains widespread and deep. Many have been "turned off," in the revealing term they use. Certainly the romance is gone—the romance that existed when John Kennedy spoke stirringly of bearing any burden in support of the cause of freedom. Attention, when given to questions of national security, has tended in large measure to take the form of attack on what is being done—attack for its costs, for its manpower burdens, for the risks involved in deploying nuclear weapons abroad, for the extent of our involvement in foreign problems, for the disagreements that occur with our allies, for the sale of U.S. arms and equipment abroad, and for much more that is unwelcome on the "cost" side of the security ledgers. At the same time, however, the benefits that these same security efforts afford—improved prospects for peace, continued safety for our people and for our commerce, freedom from outside coercion—tend to be taken for granted. When oil is cut off, or Israel threatened, or an American ship seized on the high seas, the feeling that such things should not be allowed to happen is conspicuously sharp and strong. What is especially significant in this and relevant to the present study, is the weakening of the sense as it formerly existed that peace and good world order are linked with and must be earned and re-earned with sacrifice, skill, and sustained effort.

A careful and convincing reappraisal and clarification of security and defense issues may therefore have an especially valuable role to play in this situation. The confusion and disbelief that have been in evidence may mean that events and conditions that do not conform to patterns familiar in the past have been too long presented in ways that only heighten their confusing, disturbing, and frustrating effects. If so, there can be little wonder that people should be turned off and that the mood of withdrawal should be so strong.

Many people have found it hard to understand why the United States, after several years of detente with the Soviet Union, and after supporting defense budgets now running over $100 billion a year, should have to be seriously concerned about security issues—or should see unmistakable evidence in Angola and elsewhere that all is not well in terms of our foreign interests.

Yet such is assuredly the case, and there is no sign that the need for coherent, widely supported policy and action in the international sphere will disappear in a world that, like it or not, is increasingly interdependent and increasingly vulnerable. In fact, the problems of security that concern us may well be more, not less, in the time ahead, and their complexity is likely to increase. There are grounds for the current uneasiness that events may be moving out of control, that many of our friends are in deepening trouble, that perhaps the "correlation of world forces" may indeed be moving in favor of the Soviet communists, and that the future—if we do not act wisely and effectively—may bring significant reverses for U.S. interests and influence.

The Constraints of Practical Possibility

The internal stresses and difficulties we have experienced, when measured against the concerns we must feel, have brought a sharp reminder that the security interests and objectives we set for ourselves must be *dimensioned* to be meaningful and to serve our purposes. That is, they must strike a balance between what we would like to have, and what we will be willing to support and be able to achieve. Our security interests must be measured against the practical possibilities open to us—such things as our resources, our skill, our will, and our solidarity.

With regard to our resources—both our natural resources and the industrial and commercial development that makes the United States the rich and productive nation that it is—and with regard to the skills of production, the skills of government, and the military skills we have generated, it is obvious that the United States still holds a very high rating and can continue to do so. It is rather in the area of will and solidarity that the more important factors of concern to future security are likely to be encountered. The internal divisions within the United States, the low levels of national will and morale, the lack of national confidence in our ability to cope with problems at home and abroad—all these necessarily have a constraining effect on the needs and interests that we can effectively pursue in the international arena. Any full-blown attempt to identify the causes would take us far afield, but these problems are serious and important, and their causes undoubtedly are quite profound. It must be acknowledged in this connection that past excesses have contributed powerfully to these unhealthy and disabling attitudes, together with past failure to assure—and to persuade—that our security policies, our military undertakings, and our military programs, as well as our asserted security interests, were validly related to the real concerns of our people.

A second limitation has emerged in the erosive tendencies, mentioned earlier, in the strength of the NATO and other alliances that were created in the early postwar period—tendencies that seem likely to continue because their causes are likely to persist. In large measure, this is a natural process to expect. As the apparent imminence of military threat has receded, our allies, like ourselves, have tended to turn much of their attention to other things, and to put a lower premium on the importance of continued allied solidarity. The results have included a progressive decrease in the support that can be gained for maintaining adequate military strength, a trend toward unilateral and bilateral patterns of policy, and a tendency as well to give way to old quarrels and to the things that divide and weaken us rather than those that draw us together and strengthen us. It is not clear whether and where trends such as these will begin to level off or reverse themselves. Thus far they have given little sign of doing so.

All this is at a time when the Soviets have become able to match much of what we do, and in some cases, to surpass us. They still stand far behind the West in resources and in skill—particularly in the productive skills of their population—but their levels of activity continue to rise. With regard to will and solidarity, while they may perhaps in the future undergo serious difficulties as more Solzhenitsyns, Amalriks, and Sakharovs appear among them, their effective solidarity and the will that they can place behind the decisions of their political leadership remain for the present period strong indeed.

The communists have one special skill of which we may well take note—the science they have made of knowing Western soft spots and how to exploit them. The North Vietnamese, for example, deliberately chose American morale as their prime target during the Vietnam war, just as they had chosen French morale as a target in previous times. Fortified by theories of protracted war, they relied on weariness within the United States to divide and weaken our society, and they were proved right in what they anticipated. They were skillful in exploiting some of the most powerful American institutions—the press and television in particular—to attack our country and what it was doing. They used the Asian technique of ju-jitsu to cause us to employ our own strength to defeat ourselves. But it has to be recognized, along with this, that the North Vietnamese were willing to take the pain, to undergo the loss of life, and to bear the heavy burdens and hardships that their part in this war involved, in order to carry on this prolonged conflict—and to bring it to final victory.

None of this should suggest that it is appropriate for our people to abdicate the responsibility for pursuit of our national security interests and needs or to lapse into a kind of defeatism in the face of continued dynamic communist efforts at expansion. What it emphasizes instead is—above all—the need in our country for the government—Executive and Congress alike—to provide more effective leadership in building understanding in our people of their true and valid security interests, along with a sense of the limits of practical feasibility. The internal difficulties that have been experienced highlight as well the need for top leaders to make certain of public support and to avoid taking on unnecessary tasks—in particular, unpleasant tasks that are likely to prove beyond the limits of what our people will be prepared to support when the going gets tough.

The Need for Selectivity—Two Tests of Leverage

To deal with the wide range of possible dangers we face, varying as they do in their severity and importance, at a time of significant internal weakness, we must be selective—we must practice discrimination in the burdens that we assume and the security interests that we adopt as action-goals. It is necessary to identify, with care and precision, those international events and conditions abroad that our people are likely to take seriously in terms of their security and their well-being. We have seen the harsh results of overextension—extension of our commitments beyond the willingness of our people to sustain them—in the debacle that finally occurred in Vietnam. At the same time, we cannot afford to forget that we have likewise been warned about the harsh dangers of complacency and about taking for granted the kind of favorable world order that permits steady access to raw materials and the flow of international trade by the use of oil as a weapon against us by the Arabs in 1973 and 1974.

In being selective there are two broad tests of leverage that seem useful for determining what undertakings the United States is likely to be able safely to assume, and what undertakings the United States should avoid. The first is to ask just how much real impact the particular events and conditions that may be under consideration actually will have on things that Americans care about—that is, on the things that they will care enough about to make significant sacrifices to safeguard. We learned to our sorrow in Vietnam, after we were deeply engaged, that they simply did not care that much about that far-distant people struggling to maintain their freedom. The question that remains is what will happen if the United States does not act, and how important such an outcome will really be.

This is an area in which the quality and effectiveness of national leadership can be of tremendous significance in expressing and explaining the importance of such foreign events and conditions to American security and well-being. Even so, there are limits in what can be done, as the Vietnam experience clearly showed. The intensity of impact that Americans will actually feel in specific cases is quite hard to anticipate with any precision, since the factors are so many and diverse, and their balancing is so uncertain. Nevertheless, the more important factors are likely to include such considerations as geographical closeness (as of Canada or Western Europe), national origins and emotional ties (with Italy and Israel, for example), immediate and acute consequences (oil and bauxite, for example), visible and substantial threat (Soviet missiles in Cuba), direct and certain versus indirect and ambiguous effects (the "domino" theory in Southeast Asia, for example), likelihood of quick success with limited commitment and cost (the Dominican intervention in 1965).

A classic but tragic appeal to such factors as geography, immediacy, hereditary ties, and familiarity was given in Neville Chamberlain's famous pre-World War II claim that the British had gained "peace in our time" through the abandonment of the Czechs as a "far-away people of whom we know very little." Hitler's subsequent initiatives showed all too clearly the long-term costs and hazards that may result and should suggest to us what the consequences might be of too narrow and short-range an appraisal of the potential impact of foreign events on American life. Nevertheless, this first test of leverage retains its fundamental importance for the setting of American security goals.

The second test of leverage, no less important than the first, is how much actual impact the United States can have on these particular events and conditions abroad, within the limits of the resources that the American people are likely to make available and the efforts they are likely to sustain. The skill with which such resources are used will of course influence the outcome, but three decades of postwar experience and many postwar disappointments suggest that we have entered a time in which we will lower our sights and keep them lower. We have learned abroad as at home that efforts even of large scale may prove unequal to remedying or to altering economic and political problems that are deeply rooted in national history, culture, and geography. And we know that our opponents can be just as selective and as skillful as we ourselves and particularly can choose our soft spots as targets for their attack. The contrast is sharp between an area such as Europe, where a prosperous, proven culture was being *restored* after the war and one like South Korea or Southeast Asia, where intractable problems of age-old poverty, religious strife, and lack of governing institutions were to be overcome and new patterns of national life achieved.

The two tests of leverage mean simply that the interests on which we base our security policies and build our security forces should be held to those that keep costs and sacrifices commensurate with expected benefits. The value of intervention could be calculated with greater confidence if all factors were subject to our sole control. But the essence of the international struggle is that multiple players capable of independent decisions are participating. Only within limits can we estimate what their assessments of values will be. A nation's leaders are hard-pressed to know in advance just what priority their own citizenry will place on the balance of sacrifice and benefit to particular issues. Moreover, judgments regarding interests are by no means fixed for all time. A special word is perhaps needed regarding the important matter of attitude formation in this connection. Obviously, the attitudes our people are likely to take in such matters are not apt to be wholly independent of the objective facts of the situation involved or of rational analysis as to the likely consequences of particular forms of action or inaction. But the formation of such attitudes on their part is, in addition, powerfully affected by a whole complex of subtle processes, some rational, many impressionistic and emotional. In particular, their exposure to opinion-molding influences that exert heavy control over the identification of issues, over the intensity and impact with which they are presented, and over the "line" to be put across appears to intrude more deeply into their attitude development. The role of modern television in this regard as a cost-driven, highly commercialized medium that lives on dramatic impact, high emotional content, and selectivity of subject matter, is especially potent, and it is only now even beginning to be understood. The lead time for developing deeper, more reasoned, more balanced judgments on complex and trying issues is long, and the resources for doing so are all too limited, when we recall, for example, how much television-created perceptions dominate the press, academic attitudes, and election-hungry politicians. When perceptions open to manipulation in this way form the starting premise, the risks become great that opinions will fall prey to stereotypes divorced from hard complex reality. The kind of reflection and retrospection needed are denied by the fleeting nature of the presentation, by the difficulty of (and resistance to) postaudit of editorial positions, and by the continuing shift of dramatic interest to new issues. In this process public opinion is made sensitive to a security issue, often through selective faultfinding, is shaped with regard to the issue, and then is led on to other things, rather than being assisted to develop a fuller, more balanced understanding of the issue.

Since television is and will remain a major fact of life on the American scene, and since its effects are so powerfully shaped by its inherent characteristics, its impact must now be accorded high rank as a major factor in American security affairs. Like the "military-industrial complex," the weight of its role is now so great as to deserve particular attention by the American people and their leading evaluative institutions.

A further special factor affecting American attitudes is the duration of a particular issue or challenge. Part of the extra dimension of American achievement has traditionally come from enthusiasm and solidarity. Someone has said, even before the Vietnam experience, that America should never fight a Seven Years' War. Americans tend to want not only a clear and complete solution, but a quick solution as well.

But there is nevertheless a danger with selectivity—the danger that it may result in action inadequate to maintain international order and stable security and thereby concede to a more vigorous opponent the power to shift the international balance to his advantage. Thus, while on grounds of realism, a very low posture of policy may be appropriate (and a high degree of skepticism toward new undertakings and initiatives), the result may easily be a steady loss of position and influence. The objective facts cannot be wholly disregarded, after all. And there is need to knit the selective interests once identified into a coherent, viable whole, carefully related to the strategic and security environment in which these interests are to be pursued.

2 The Strategic and Security Environment

Of the multitude of security factors—variable or constant—that condition American security prospects for the years ahead and shape our security interests, three trends have been of special significance in recent years and seem likely to continue to be so.

The first is the overall shift in the balance of military power in favor of the Soviet Union that has been evidenced since broad-scale Soviet military expansion was initiated in the mid-1960s, an expansion that stands in clear contrast to the continuing reductions and curtailments in recent years in United States and other Western forces. The prospect is for a continued shift in the same direction to occur in the years ahead by virtue of the momentum demonstrated by the Soviet activity, especially the military research and development program, which has been proceeding at a rate estimated to exceed that of the United States for a decade or more. The Soviet urge to acquire and maintain massive military strength quite obviously responds to something deeply and historically imbedded in the Soviet mentalities as well as the calculating Soviet assessment of possible specific use and role of military power in the international arena today. And on the Western side, the downward drift is in part a reflection of careful calculation of force programs and budgets but in large part—probably in greater part—is the by-product of a shift of concern and interest to other demands and proposals, most often of domestic rather than international nature.

A second trend powerfully at work is the already substantial and steadily increasing dependence of the United States and other Western industrial nations on raw materials from abroad. For the ores of more than a dozen important metals essential to our industry—including manganese, bauxite, tin, chromium and tungsten, among others—the United States currently imports more than 50 percent of its total consumption. Foreign oil imports, less than 20 percent of total U.S. consumption at the time of the Middle East war in October 1973, less than three years later had risen to more than 30 percent—and this despite intense public and political discussion about the need to *reduce,* rather than increase, such dependence. The inclusion in the list of source areas for such ores and energy supplies of many countries located in deeply troubled areas of Southeast Asia, southern Africa, and the Middle East makes clear the strategic hold that this dependence—now an established international fact of life—has come to exert upon the United States. And the USSR itself is a substantial supplier for U.S. consumption of chromium, platinum, nickel, and germanium-indium.

The third trend—much less clear than the other two—relates to the present

21

and prospective state of world order—the ability of international institutions and supporting arrangements to safeguard the peaceful conduct of international affairs. Here the security picture that confronts the United States is decidedly mixed.

One view accentuates the negative: the long-established patterns of world order and security, maintained in the past chiefly by Western predominance, are giving way to new influences. The phenomenal advances of Western Europe, America, and Japan—economic, financial, intellectual, and cultural—have been slowed, stopped, or matched elsewhere. The collapse of Western European power following the bloody losses of two world wars, the defeat of Japan, the end of the colonial system, the decline of *elan vital* first in Western Europe and then in the United States, the limited effectiveness of the United Nations, all have joined to end an old security order without creating a new one. Competitive short-term self-interest and the politics of xenophobia characterize much of the international scene. The United States proved unwilling to bear the burdens of leadership in evoking common effort and common purpose, sacrificing its Vietnam ally in a war already effectively won. The oil weapon in the hands of the Arabs showed how fragile Western unity had become, and propelled Italy and Britain, already in deep economic and financial distress, into a further downward slide. The system of currency relationships broke down, without, it is true, triggering the "beggar-my-neighbor" consequences of the 1930s, but with a nonetheless severe setback to Western industrial and commercial growth. The United Nations—except for a handful of successful (and valuable) peace-keeping efforts—not only failed to provide institutions to relieve the stresses and dangers but became in fact a mechanism to crystallize and then to aggravate the confrontation between "have" and "have-not" nations and to sharpen rather than ease the threat to the existence of Israel.

And finally, to the extent that free processes of government serve to produce greater well-being, together with a desire for peace and respect for other peoples, the continuing slide of noncommunist nations into authoritarian rule steadily darkens the prospects for international order and concord.

The outward thrust of Soviet communism has provided a means to exploit precisely such tendencies—the support of "just wars of liberation" (less euphemistically, the takeover of noncommunist nations by communist regimes allied with Moscow). There are power, conviction, and an impressive body of technique and experience behind the Soviet thrust that are not matched or counterbalanced by the West. The spreading presence of Soviet military power, the key base areas to which they have been gaining operational access, and the wider deployments in greater strength of military forces that they continue to modernize and expand offer them an increasing capacity to challenge a world order committed to peace and national independence. And as strategic nuclear weapons are brought under stricter constraints, the Soviets gain a freer hand to follow just such a course. While they have preferred—in Greece, Korea, and

Vietnam, for example—to seek their goals by proxy insofar as actual combat is concerned, their readiness to intervene with massive support has been well demonstrated in Nigeria, Cuba, and Angola. The possibility of their intervention reached the high point of threatening the introduction of airborne troops into the combat zone during the Arab-Israeli war of 1973—a move that recalled their proposal in 1956 that they join with the United States in deploying troops into the same area at the time of the Suez crisis (a proposal that was quickly rejected by the United States).

The trends adverse to world order and stable peace are thus weighty and substantial. Yet a persuasive counter, admittedly less than conclusive and complete, also can be argued. On many important security issues, progress in lessening the likelihood of large-scale war has been impressive. Particularly since the initiatives of President Nixon, direct channels for the presentation and resolution of issues between the United States and the USSR, and between the United States and China have been opened and deepened, and negotiation has been substituted for confrontation. The SALT agreements reached to date, and the continuing search for others, promise some respite in the increase of nuclear missile levels and advance in weapons technology. NATO, whatever its needs for refurbishing and revitalizing, remains a cornerstone of Western security architecture. The relationship of the United States with Japan and South Korea fulfills a comparable role and purpose in that area, providing a secure northern anchor for peace, stability, and efforts of mutual benefit in the western Pacific.

Even in the turbulent world of the developing countries recently freed from colonialism, a number of strong and favorable trends can be discerned. The most striking among them is the gripping power of the idea of sovereignty and national independence. In Latin America, Africa, the Middle East, and much of Asia, the strength of nationalism is a dominating and directing force. Whatever the abrasion with the West that results, or the occasional lapses into chauvinism, the nationalistic fervor has thus far been strong enough to forestall any twentieth century renewal of the kinds of land grabs by the great outside powers that were seen repeatedly in the past. Where put to the test, the sense of sovereignty has proved its strength: in the rejection by the Arab countries of Soviet domination despite their intense need for Soviet help, in the Congo (Zaire) and Nigerian attainments of their own internal solutions, in the rejection by many others of hegemonic dominance or control by the USSR or China. Although there are significant exceptions, as in Soviet military presence and influence in Somalia in Eastern Africa and at Conakry in the West, nationalism has normally carried the day. Whether its strength has come from a feeling of national identity, or from a conviction that separatism would be disastrous in its result, the nationalism that has been demonstrated has held off outside domination and—particularly in Africa—has prevented the dismemberment of national units, whatever the intertribal stresses or the initiation of attempts by force to change existing borders, however arbitrary they often have been. The obvious exception in South Asia—Bangladesh—has not been regarded as a model to emulate, but as the unhappy product of Indian-Pakistani animosity.

Nor has the rise to independence of the new countries simply generated a destructive struggle of each against all. Within regional groupings, frail and fragmentary as they often are, the mechanisms that have been created are achieving a measure of common action and identity and mutually advantageous cooperation. Moreover, many of the former colonies have built new relations with the former metropole—the French colonies with France, and Zaire with Belgium, for example—that are sounder and healthier for having been built on the basis of legal sovereign equality.

In key aspects of international behavior, there has been slow but noticeable movement toward improved standards of conduct. There is a growing emphasis on trade (even though accompanied by heated debate over the terms of trade), on international finance, on migration of labor and economic or commercial activity of many kinds, on travel and education abroad, on cultural and intellectual exchange, and on ideological competition rather than military attack. The world noticeably has been turning away from the use of force—at least from the large-scale use of force. Border wars are increasingly viewed as an anachronism if not an absurdity. Except for internal wars (many of them a final stage in the process of decolonization), the list of armed clashes tends to narrow to those few hard cases that have unique historical antecedents—the Arab-Israeli struggle, the Indo-Pakistan quarrel, the clash of Greek and Turk communities in Cyprus, conflict between Arab and Christian communities in Lebanon, the lingering Kurdish resistance in Iraq, and the conflict of Catholic and Protestant communities in Northern Ireland. It remains a possibility of course that in divided countries—Korea and Germany, specifically—an effort will be made, in all likelihood from the Communist side, to reunite the country by force. But even there, it seems more likely that the communists will continue to pursue processes of infiltration, subversion, and takeover from within.

Likewise, in the longer term, the possibility must also be recognized of disintegration of the Soviet "Empire" itself—initially at least through the breakaway of the nations of Eastern Europe. But the processes by which such events might be brought to occur would inevitably be of such politically cataclysmic proportions as to open a whole new security era—undoubtedly a perilous one—in international affairs.

Between these two views—of a breakdown of world order, or, alternatively, of the growth of a successor better suited to our times—it is not possible to make a confident black-or-white judgment. Both tendencies are unquestionably at work, in differing ways in different parts of the world, with differing effects on American security. It is appropriate then to take a somewhat closer look in turn at such major distinctive areas as the Soviet Union; Western Europe; the North Atlantic and North America, the Pacific, Asia, and the Far East; Southeast Asia; the Middle East; Africa; and South America. This look will identify the chief considerations that are of concern for the purpose of this book, and will identify in particular those considerations that relate to American security interests, security policies, and the role and structure of the armed forces and the military programs that support them.

The Soviet Union

The Soviet Union, together with adjoining states under its political hegemony and military dominance, today and for the foreseeable future, holds a dominating position and significance among the principal regions of the world influencing American security. Its industrial strength and military power unquestionably place it, with the United States, in the first rank of nations for the remainder of this century. It is the major military and political challenge with which the United States must deal, and barring internal upheaval it will surely continue in this rank, although any major erosion of its control over its Warsaw Pact partners in Eastern Europe would of course detract from its freedom of action. Its massive military forces—including nuclear ballistic missiles aggregating many thousand megatons in their destructive power—pose a constant and continuing threat to every other major power center of the world, including the United States and Western Europe, as well as Japan and China. While the capability of the Soviet armed forces to project large-scale operations beyond the land limits of Europe and Asia is still limited, their capacity to do so is growing, and the processes by which the Soviets expand their presence in peacetime circumstances are being tested, demonstrated, and strengthened in Africa, South Asia, and the Middle East.

Committed to the continued expansion of the communist system—although without, apparently, a fixed timetable—the USSR presents itself, in the world arena, as an "unsatisfied" power with active ambitions hazardous to the health and security of noncommunist nations and dangerous, as well, to any concept of world order based on the principle of freedom of nations from outside threat and coercion. While the strength of its powerful neighbors in Western Europe, Japan, and China—backed by their ties to the United States—has been sufficient generally to hold it in check, the Soviet Union's outthrust and outreach remain a dominating feature of the world security environment. The Soviet Union continues as the primary source of, organizer of, and exploiter of the threat to the security and stability of other nations and other regions.

Nevertheless, a number of welcome developments have occurred that have eased the security problem for the West from the more acute and dangerous stages of the past. Soviet willingness to negotiate with apparently serious intent has brought improvement, via the SALT agreements and initiatives, in the strategic nuclear confrontation. The Mutual and Balanced Force Reductions (MBFR) negotiations offer comparable possibility for stabilizing the military situation in Western Europe. And past flash-points such as Berlin seem to have been considerably defused through the agreements finally concluded in the last few years. The Soviet forces positioned from East Germany and Poland to Czechoslovakia and Hungary give a depth of protection to the Soviet borders that undoubtedly eases Soviet concern as well as Soviet truculence. The enforced division of Germany likewise spares them concern for the present over the possible future resurgence of the formidable adversary they have known there in the past.

The tone of international discourse on matters of security has certainly improved, and the subjects on which serious discussions are held—from nuclear weaponry to crisis consultation, and from computer technology to cultural exchange—have certainly broadened. But even while the sense of eyeball-to-eyeball confrontation has dropped away, Soviet military power has continued to grow in magnitude, and its military technology has advanced on every front— Soviet rocket forces with whole new families of ICBMs, the Soviet navy with aircraft carriers of large size in operation for the first time, the Soviet ground forces with new self-propelled artillery and mechanized infantry vehicles, the tactical air forces with advanced aircraft for tactical air support, the air defense forces with new classes of surface-to-air missiles. All this is backed, as noted, by a military research and development effort exceeding that of the United States, all the more significant because the United States has heretofore been accustomed to rely on "quality," i.e., superior advanced weapons technology, to offset inferiority in numbers. The Soviet military forces have become the greatest concentration of military power the world has ever seen, far exceeding any reasonable estimate of what would be required for the defense and security of the Soviet Union itself—or for the defense and security of the Soviet Union and its present allies. In nearly every category of armed force except aircraft carriers they by themselves exceed in strength the forces of the United States, and they with their allies exceed the total strength of the United States and its allies. In overall terms—although there is room for argument over particular detailed force categories—they have now gone well past the United States in aggregate military power.

They have joined with the United States in the constructive work of the SALT negotiations as a step of worldwide and historic importance to place ceilings on strategic nuclear weapons and to place restraints, still limited to be sure, on destabilizing influences and the development of crisis conditions that might lead to nuclear war.

But they still press forward with efforts on many other fronts to extend their presence, their influence, and their hegemony in further areas of the world. In Africa and the Middle East, their military power forms an ever-present backdrop to a wide range of efforts—whether to sustain their position and connection in Syria and Iraq, to install a regime of their choosing in Angola, or to establish bases and base rights in Conakry and Somalia. In Vietnam they showed their readiness to continue to furnish military aid to an active foe of the United States, even after the United States-North Vietnamese agreements of 1973 were concluded, and in violation of those agreements. Using the Cuban military as their *corps d'intervention*, they established a presence in the heart of Africa—through the Angolan enterprise—capable of employment for further political-military objectives in every direction of the compass.

We see then a reduction, carefully limited in its scope and impact, in international tensions (and even in some of the possible *causes* of tensions, such as a continuing expansion of nuclear missile forces), but we do not see by any means an elimination of such tensions. A carefully orchestrated policy toward the United States permits the Soviets to avoid actual confrontation while continuing to pose a nuclear threat to the world around them and while probing for opportunities to expand and exploit the international openings created by crisis or conflict. They join with the West in stabilizing and legitimatizing existing borders in Central Europe (together with their peacetime military presence in Eastern Europe), but they keep continued military pressure on Western Europe and the Atlantic area.

While the security relationship between the USSR and the United States is in large measure bipolar in nature, the three-cornered relationship of the USSR, China, and the United States has also strongly shaped the world security balance, particularly since the breach between the USSR and China became visible to the West in the early 1960s. The mutual concern and distrust between the Chinese and the Soviets are deeply rooted in historical conflict, present disputes, and the prospect of future clashes sparked by conflicting worldwide ambitions. Militarily, each has taken a largely defensive posture toward the other; the substantial size of the respective opposing forces in Asia shows the measure of their mutual concern. While the pattern of the dispute has been largely one of polemics, of propaganda beamed into border regions, and of struggles within the world communist movement, military clashes have occurred on occasion in the past and could occur again. Meanwhile, China loses no opportunity to warn and encourage NATO to keep up its strength on the USSR's western borders, and thus keep alive in Soviet minds the specter of a two-front war.

The relationship between these two countries is highly significant to U.S. security interests and warrants constant observation. A major effect is the restraining influence it tends to exert on both China and the Soviet Union. If either country were to get seriously involved with the United States, it would at once become vulnerable to pressure by the other. And if the two were to become militarily embroiled with each other each one would surely fear that the power of the United States could tip the scales against it.

The USSR and Chinese memories of past Soviet conquests in Asia and periodic earlier incursions from Asia westward, when added to present-day Chinese concern over Sinkiang and Russia's uneasiness over its Central Asian provinces, will surely continue to color their relations. While they may moderate their policies in order to lower the level of vituperation they have at times reached in the past, it is hard to see how any coalition of *entente* between them could prove stable for very long. For two neighbors of such power, messianic commitment, and heritage of suspicion, the tendency for one to dominate or subvert the other, and for each to suspect the other, seems sure to prove upsetting before too long. While militarily this situation is nothing the United States has any interest in trying to exploit actively, the restraint that is enforced on both China and the USSR in their relations with the United States by their own quarrel is a welcome alternative to the possibility of joint communist militancy. But because Chinese-Soviet reconciliation, however transitory and unstable, can never be wholly excluded, the U.S. need for an adequately strong deterrent posture against each must not be forgotten in the moments when their quarrel becomes intense.

Even in the case of the bipolar U.S.-USSR relationships and interactions, the interests of other nations, such as the NATO partners of the United States, cannot be overlooked. In the course of the confrontation between the two during the Cuban missile crisis in 1962, the dangers to Europe were keenly felt by the NATO Allies. Later, when the U.S.-USSR agreement for consultation to avoid nuclear war was signed in 1973 there was sharp European concern whether the U.S. nuclear umbrella was somehow being removed, or its coverage curtailed. And though the SALT negotiations were from the outset bilateral, the U.S. wisely took special measures to keep its allies informed, and to confirm to them periodically that the negotiations were continuing carefully to exclude any weapons issues that involved the forces of the other NATO nations. Even so, grounds for concern could not be wholly removed, since the U.S. strategic nuclear force inevitably remains the ultimate deterrent in support of NATO's collective defense posture in Europe, and thus, for the NATO nations, the final safeguard of survival itself.

Western Europe

The region of Western Europe, whatever the turbulent and changing conditions elsewhere, remains a major center of world power and a mainstay of Western security. Its strategic importance to the United States is of the highest order. The ties that exist, despite points of difference and occasional disarray, are strong and rewarding to both sides—and they extend over a full range of political, economic, and military cooperation. Modern thriving industry, advanced social systems, and flexible and resilient governmental institutions place Europe, even in its postcolonial phase, in the forefront of world affairs.

Its stability, security, and confidence are tightly bound up with those of the United States. Lying in the shadow of Soviet military power, which stands ever ready to give support to efforts from without and within to place communist regimes in power in the western European countries, these countries have need for the support and assurance their continued close association with the United States gives them—an association given reality by the presence of substantial U.S. forces in Europe and by the commitment of these American forces to the common NATO defense goal.

The existence of NATO—whose institutions unquestionably represent one of the great constructive achievements of the present war-ravaged century and constitute a continuing major contribution to Western strength and security—has proved its value to Western Europe and North America alike. While Western Europe evolves slowly and haltingly toward organic unity, NATO continues to provide a shield safeguarding its independence and its freedom of action.

There are, nevertheless, numerous evidences of deterioration and weakness within Western Europe, and NATO itself, that provide grounds for concern as we look to the future. Economic stresses and the acute vulnerability of the oil supplies on which Western Europe is so vitally dependent have caused divisions in the alliance and could do so again at any time almost without notice. Quarrels based on ancient disputes and grievances can be reignited—by events such as Cyprus, for example—and do deep damage to the alliance. A general weakening of allied security in NATO's southern region has occurred in recent years chiefly as the result of such issues and events. Politically volatile conditions exist throughout the length of the Mediterranean, ranging from the governmental transitions that have been occurring in Portugal and Spain, the troubled economic and political conditions in Italy, and the uncertainties that will follow the passage of Tito, to the continuing dispute between Greece and Turkey.

A slackening of defense effort, a loss of support in several European countries for defense programs and needs, and even a weakening in the will and determination to stand firm against the Soviet Union are further evidences of decline. Looking back it is evident that the negotiations of the West with the Soviets in the early 1970s contributed to a degree of relaxation and complacency (reflecting a considerable shift in psychology arising from detente) that was quite unwarranted by any actual change in the fundamental Soviet policy or in its steadily increasing military power. By 1975, experience with the hard Soviet positions taken in CSCE and MBFR negotiations, viewed against the background of their continued military strength, had restored in the NATO countries a more cautious outlook (shown in particular in the firm, thorough preparations on the NATO side for the MBFR negotiations).

In military and geographical terms, Western Europe lies across the path of any attempted Soviet advance to the open Atlantic and the Mediterranean. Should war occur, these territories would of course be the scene of heavy fighting. But the organized collective force that is in place, by just being in Western Europe, provides a powerful deterrent to any thoughts of initiating such a war on the part of the nations of the Warsaw Pact.

The opposing blocs formed by the countries of NATO and the Warsaw Pact give an essentially bilateral character to the security issue in Europe. Nevertheless, it would be an oversimplification to see it in that light alone, since both blocs are far from monolithic and are affected by the distinct interests and activities of the individual members. While we know far less about the internal limits and differences that exist in the Warsaw Pact than about those in NATO, it is nonetheless evident that they are significant. Rumanian resistance to the presence of foreign troops, and reported purges of officers of uncertain reliability in Czechoslovakia, to cite but two examples, inevitably leave questions as to the unity the Soviets could expect to find, particularly for any military adventures in Western Europe.

And no one who knows NATO well could fail to recognize its internal divergences, which are part of the warp and woof of the alliance's activities, both military and civilian. The search for common policies and the development of common plans and programs form a central feature of the daily life of NATO at every echelon.

Nevertheless, the close relationship between the United States and its allies is unique in its significance. In the quarter century since World War II the alliance has provided a true cornerstone of collective security, and it continues in the final quarter-century now beginning. For each of the Western nations, the alliance relationship that it enjoys with its partners enables it to stand up against the outward thrust of Soviet power.

The North Atlantic and North America

Our deep and vital ties to Europe give the North Atlantic much of its strategic importance to U.S. security. A dense network of trade routes carries much of the shipping on which American and European industry both depend—routes extending across the Atlantic itself, into and through the Mediterranean, through the Caribbean, and around Africa. To economies based on oil, any interruption would have immediate and critical effects.

Likewise, in war the North Atlantic would retain its preeminent importance for the passage of reinforcements and logistical supplies to provide sustained support for combat operations. Undoubtedly Soviet forces—its submarine fleet, deploying many scores of undersea units, in particular—would seek to disrupt such supply, and the North Atlantic would quickly become one of the major arenas of such a conflict.

It would be, as well, a launching area for submarine-fired ballistic missiles (although the increased ranges for the newest classes of such missiles—in the 4,000-mile range—now lessen the need to deploy into forward positions near to the enemy homeland). Substantial allied naval operations in the Norwegian Sea would be anticipated, both to support Norway and to blunt Soviet attempts to project naval power westward around the North Cape. Iceland and the Azores, by virtue of their locations, would be strategically important for antisubmarine warfare, surveillance, and the staging of forces transiting the Atlantic. Particular significance would attach to the Greenland-Iceland-UK gap through which submarines of both sides would be endeavoring to operate. Maritime operations in the Atlantic area would be of many kinds, working from islands and other base areas, protecting the islands from Soviet use and attack, defending the North American continent, and safeguarding Latin America and the approaches to the Panama Canal.

The North American continent itself, in addition to constituting a major source of warmaking power, would serve as a base for the launching of ICBM operations and for the conduct of air defense. For U.S. and Canadian defense operations, the continent has long been viewed as forming a military and geographical entity; cooperative military arrangements well recognize the close identity of defense interests and needs. The BMEWs (ballistic missile early warning system) sites located in Greenland and Alaska support Western long-range radars, which can give at least a few minutes advance warning of bomber or ICBM attack from the Soviet homeland.

The Pacific, Asia, and the Far East

Further to the west, Asia and the Pacific present a strategic significance to American security that, although demonstrated in hard experience in World War II and in the Korean and Vietnamese conflicts, is still marked in important areas by questions and controversies. Of its overall importance in world affairs, and to U.S. interests, there can be no doubt. More than half of all humanity is located there, together with vast economic resources, agricultural activity, and, in the more advanced areas of Japan, Taiwan, and South Korea, modern industrial complexes. Japan, China, and the USSR abut as neighbors, the United States has major interests and a wide range of contacts, and Europe also has extensive commercial and historical ties. In the Northeast Asia/Northwest Pacific area, the cooperation that has been developed involving the United States, Japan, and Korea has proved to be a major source of stability, security, and opportunity for economic progress.

By its location, wealth, and strength in every department of human affairs, Japan provides a stable and impressive foundation for security arrangements in the area. As our most important ally in the whole Pacific region, Japan exerts a stabilizing and constructive influence throughout Asia and the Western Pacific. Japanese military forces, although limited, nevertheless afford a significant capability for defense of the home islands and for deterrence of attack or invasion. The bases made available by Japan for American forces have been essential to the effective contribution that this crucial American military increment makes to deterrence and stability in the area—including, in particular, the nuclear umbrella that the United States furnishes in the absence of any Japanese nuclear forces or weapons. Operating from these bases, the United States is able to maintain a presence—chiefly naval and air—that gives protection to shipping routes essential to trade and industry. Japan, in turn, is able to derive confidence and added influence in Asia from the American guarantee.

Korea retains its key strategic importance, both to the security of Japan and to the maintenance of stability and Western strength in the general area. With its powerful battle-proved commitment to continued independence and its unflagging resistance to communist threat and pressure, South Korea is positioned to impede and prevent Soviet or Chinese expansion, whether through peacetime encroachment or overt North Korean attack. The possibility of renewed hostilities is, of course, ever-present. However, so long as the American presence and commitment continue—supported by bases and facilities essential to such presence—prospects for stability are enhanced.

South Korea presents the classic issue of American attitudes and policies toward an ally that is authoritarian in its mode of government but stalwart and strong in a strategic area. The manner in which U.S. public and government—both Executive and Legislative—will resolve the issue cannot be foreseen with confidence. The uncertainty will remain a source of uneasiness and possible tension, and much will probably depend on individual instances or minor aspects and events.

The future role of mainland China as it bears on American security is far from clear, and it is important to remind ourselves both how much is unknown in this regard, and how impact-laden the future course of Chinese policy can be. The basic facts are impressive. The Chinese population of 800 million, a viable economy—largely agricultural but with an expanding, already substantial industrial segment—and vast natural resources including coal and oil give China the ability to dominate the Asian mainland. While tangible results from the economic efforts being made are greatly impeded by the deadweight of poverty and the needs of the massive population, technological advances are nevertheless being achieved. Chinese rivalry with the USSR constitutes a deep-rooted and dangerous source of continuing potential instability that, however, has tended thus far to impose a considerable degree of restraint on both countries insofar as external adventures are concerned. Hence the prospect of overt conflict by either with the West in the Western Pacific area is decreased. The unresolved issue over Taiwan and Chinese doubts as to American future policy constitute perhaps the most specific and serious source of uncertainty affecting future U.S. interests and U.S.-Chinese relationships.

Southeast Asia and the Southwest Pacific

In Southeast Asia and the Southwest Pacific, future prospects are much more obscure. A new chapter is unfolding. It is difficult to trace the prospective strategic significance of the Southeast Asian area to U.S. interests, since the latter are distinctly in a state of flux with regard to this region, and the countries of the area are subject to so much continuing internal turbulence. The area is a scene of political instability and of vulnerability to the imposition of communist regimes; it is an area that has been shown by the U.S. abandonment of South Vietnam to lie beyond the area of probable sustainable U.S. involvement. The probability of communist subversion and domination of most of or all the remaining free governments of the area must be rated high.

Within the region a few particular areas retain special strategic importance from the standpoint of future American interests. Malaysia and Indonesia, for example, stand astride the important routes of sea transit from the Pacific to the Indian Ocean. They are sources of important industrial raw materials. Bases in the Philippines enable the United States to maintain a forward presence and a capability for support of wartime operations in case of necessity. Together, these give a measure of protection to trade and to national independence of the states of the area. Australia and New Zealand provide a zone of stability and well-being, removed from the main channels of world threat and conflict; their contributions to the world's economic life are significant, particularly in raw materials. Their limited populations and detached positions suggest that they will not, however, exercise a major strategic role.

The Middle East

The Middle East has become in just a few short years an area of acute and paramount strategic and security significance, not only to the United States but to the other major power centers—the USSR, Western Europe, and Japan in particular—and to nearly all nations of lesser size as well. Its oil gives it tremendous leverage on the economies of the world, advanced and less advanced alike. So long as the unresolved Arab-Israeli conflict continues, and with it the possibility of renewed hostilities that could again bring the United States and USSR into confrontation, the Middle East will hold a place in the first rank of security problems to which U.S. security policy and effort must be directed.

Hostile control of the area could threaten the economic viability and security of the West, to whom access to its vast oil reserves is indispensable if anything like current levels of gross national production—and real income—are to be sustained. The mere possibility of a new oil embargo, or the conduct of economic warfare, against the West is deeply disturbing in its implications. The prospects are high for continued accelerated social and economic change within the area, with a high measure of latent political instability in each of the major oil-producing countries. A serious source of potential conflict exists between radical Arab states and the moderate or conservative nations of the region.

In its geographical location as well, the area exercises a major strategic role, linking Asia and Europe with Africa and dominating major shipping and air routes to the Mediterranean, into the Persian Gulf, and along the east coast of Africa.

An asset contributing to stability in the area is the mutually beneficial affiliation of the United States (and U.S. oil interests) with key states of the region—in particular, with Iran and Saudi Arabia. Nevertheless, there are serious tensions between many states of the area, the effects of which could spread rapidly. In this regard the accumulation of modern arms in the area raises the level of potential violence that would have to be expected in the event an outbreak of serious hostilities should occur. Saudi Arabia and Iran, in particular—the states most critically important to prospects for peace and stability in the area—are devoting major resources to the development of modern forces.

Africa

To the south, Africa holds a multifaceted and growing strategic significance for the United States, as it also does for its allies. The interests of the major powers—Western Europe and the United States, the Soviet Union and China—are in conflict at multiple points from the Mediterranean Arab nations to the states of the southern tip. The potential for instability, both within the existing states and among them, is as high as anywhere in the world, and it seems likely to remain so. In East Africa, deep-seated problems create instability in Ethiopia, Somalia, and Uganda. Natural resources—raw materials such as aluminum, chromium, copper, and manganese for the industries of Europe, America, and Japan—can quickly convert disturbances there into worldwide difficulties. The long-standing unresolved black-white confrontations over Rhodesia and South Africa add to the likelihood of future racial strife, and of outside communist support to indigenous radical factions and regimes; they increase the hazards for U.S. policy in determining what position to take in the quarrels that seem certain to intensify.

Geographically, African domination of access by sea and air to the Middle East, South Asia, and the Southwest Pacific from Europe and the Atlantic coasts has been steadily growing in significance. Port facilities—especially those for the long voyage around the Cape of Good Hope—are important both to naval forces and to seaborne commerce. Some 75 percent of Middle Eastern oil now transits these sea lanes.

The full pattern and extent of future Soviet involvement cannot yet be predicted; in large measure it will depend not only upon the nations of the area and the Soviets, but upon the West—and China—as well. The early trends may, however, be seen in the expanding Soviet activity and presence in Somalia (including the base at Berbera), which is a source of instability to its neighbors. Soviet use of Conakry on the West Coast, and its deep involvement in Angola further to the south have already extended its strategic position and capabilities into the South Atlantic, and the whole of sub-Saharan Africa.

The probable extent of future U.S. involvement is likewise not clear, although the trend has unmistakably been toward limiting and avoiding commitment beyond diplomatic and economic contact and consultation. Nevertheless, the hope of the Western industrial countries for continued free and peaceful use of sea routes in the African area and for access to raw materials is undiminished, even in the turbulent period that probably is ahead. The disparity between Soviet (and Chinese) activism in Africa and the Western hope for good relations with limited involvement and commitment strongly suggests that the resulting trend will prove generally favorable to communist influence in the time that lies ahead—at least to the point at which the African countries themselves, like the pro-Soviet Arab countries in the late 1960s and early 1970s, may find Soviet activities too intrusive and seek to curtail the Soviet presence and pressure. The contrary precedents, however, are still numerous and clear—quick Soviet response in the form of military assistance for Nigeria, for example, or Soviet efforts to heighten their influence in Libya and Mozambique.

The Soviet base for subversion in Africa has thus been strengthened. Africa is beset by circumstances that leave it especially susceptible to subversive techniques: unsettled political conditions, rapid and massive social changes, racial confrontation in the southern tip, and numerous tribal and border disputes that could overturn the initial African policy of deliberately avoiding being drawn into such conflicts. Authoritarian factions and regimes that are closely linked to and supported by the USSR or China provide a ready vehicle by which takeover by communist-oriented groups can be accomplished.

Central and South America

Within our own hemisphere, a special premium has attached for many years to friendly and good neighborly relations with the nations of Central and South America, for reasons of U.S. and hemispheric security and for the attainment of mutually shared economic and political benefits. While ties have often been uneasy with particular nations and on particular issues, the sense of necessity in developing and maintaining harmonious policies has been strong and widely shared. In contrast to other areas of the world this is a region that has been notably free, with few exceptions, of territorial, racial, or religious disputes carried to the point of military conflict in recent decades.

Geographically, the region is vital to U.S. security and well-being. Hostile nations, especially if linked to alien power antagonistic to the United States (the Cuba-USSR combination, for example) would pose a hazard to the United States that could become intolerable if it were to be substantially repeated. Not only the nearness of our southern neighbors, but also their positions to our flank and rear as we confront the threat inherent in the massive military deployments of the Soviet Union, intensifies the strategic significance of the region.

The opportunities that the region offers for mutually advantageous trade and for political and military cooperation in behalf of security and stabilty also are important. Two-way trade between the United States and the countries of Latin America was in excess of $10 billion a year by the mid-1970s. While domestic instability—political and economic—has been a chronic problem for many of the countries of the southern continent, generally good interstate relations have existed, even during periods of great turbulence and turmoil. Although terrorism has been making inroads into internal and hemispheric order, the governmental counteraction has been substantial and of significant effect (albeit no one could claim that the problem—in nations such as Argentina in the mid-1970s, for example—has shown convincing signs of being brought under control).

Because the economic problems are severe and widespread, they create political cleavage and struggle in many countries that give grounds for serious security concern. As a major world producer of raw materials—such as copper, bauxite, tin, and oil—South America remains a potential source of economic disturbance that would have worldwide effect. The security of shipping routes, through the Caribbean and along both coasts, and the safe, uninterrupted use of the Panama Canal are of special importance for the trade of Latin American countries with each other and with the United States (and also with Europe and Japan), since land and air transport inland are limited by terrain and distances in their role and carrying capacity.

In the coming decades, there is the clear prospect that South America will take on a growing weight in the world arena, as economic growth (especially impressive· in Brazil), expanding population, and enhanced political influence increasingly make themselves felt.

The High Seas

The strategic significance of the oceans of the world to U.S. security and well-being needs little elaboration. A tremendous and diversified trade in raw materials (including oil) and finished goods—vital to the continuance of modern economic life—remains critically dependent on free, flexible, confident use of the sea. Although the United States is somewhat less dependent than insular industrial nations such as Great Britain and Japan on ocean shipping, our dependence is fundamental to our security, and it will remain so. Militarily, the seas provide indispensable links with our allies and operating areas from which to project military attacks against an enemy in case of actual war or to defend against attacks—attacks that are fully capable of being launched against our lines of communication, or even against our own territory, employing a large and powerful fleet of missile-carrying submarines.

Not only the open oceans, but also the economical sea routes along the continental coasts and the narrow seas and constricted passages have their important strategic role to play. And in future years—already, in areas like the North Sea, the Gulf of Mexico and the U.S. continental shelf—offshore economic activity will become simultaneously a new and expanding contributor to national and international well-being and an added charge on arrangements for security and world order.

A brief survey such as this obviously cannot be complete in tracing the security implications for the United States of the circumstances and developing issues in the major areas of the world. Each of these could require a complete treatise. Nevertheless, these highlights, limited as they are, serve to identify the principal matters with which our security interests are likely to be concerned; they suggest the nature and some of the probable limits of U.S. military capabilities and dispositions that would support and advance them.

The time-span with which we concern ourselves for such an analysis is the "foreseeable" future, a term that begs the question but usefully reminds us how fallible are the human gifts of prophecy. This brief and skeletal consideration of our security environment draws to our attention major real problems and real opportunities—both existing and anticipated—that is, those that are likely to exist if we do nothing, or do no other or no more than we are now doing—or even, in some sobering cases, despite anything or everything that we could do. This latter suggests that our aim will be to consider what improvement is possible in the specifics of the world situation we will be confronting—whether in the USSR, in Europe, in Asia, or elsewhere—what improvement is likely if we do the best we can. The task is one of progressively isolating and identifying the circumstances, values, and incentives that are likely to trigger the need to take action beyond our borders—action to support, to cooperate, or to oppose. Because of the long lead time in forming and training military forces and in designing and producing the weapons with which to equip them, it is necessary both to search hard for the circumstances and trends that have enduring effect, and to ready ourselves to revise and refine these basic estimates whenever events and better knowledge so require. Thus we should lay down plans and commit resources to them on the basis of current, although imperfect and incomplete, appraisals.

3　U.S. Security Interests

In identifying a set of security interests on which to design security policy and force structure, our primary guide is the evocation of American security interests wherever and whenever our highest national values are placed significantly at risk in the international arena. But American security needs and values at risk are not the only standards that must be applied. Criteria of feasibility and sustainability must also be carefully observed. Here, past experience furnishes an especially valuable guide. It helps to tell whether we are adhering to reasonable limits of will, national effort, morale, and budgetary support; whether the combat operations that are foreseen as possibilities are militarily sound—i.e., whether they are consistent with the capabilities of the force, and free from excessive risks, including the risk of high human losses; and whether we have previously been successful in achieving stability and security at acceptable cost. In the absence of convincing alternatives, the simple maintenance of the status quo, particularly as crystallized in successful treaty commitments such as NATO, provides a useful point of departure.

The security interests that we as a nation define for ourselves should allow for a considerable prospect of change. Looking toward the 1980s, the relevant changes include a rising economic struggle between the "haves" and the "have-nots" and especially between raw materials producers and consumers. The patterns of cooperation and conflict with the United States will be constantly changing; advances in military technology will alter the balance between offense and defense (with some overall advantage probably accruing to the latter) and will add to the destructiveness of war; changes in the size, composition, deployment, and weaponry of foreign forces—friendly, unfriendly, or neutral— likely will be of particular importance in the Soviet bloc and in the oil-rich Middle East; there will be a continued lag in NATO strength and modernization compared to the efforts of the Soviet Union. Attitudes will harden toward the United States, notably on the part of third-world countries sliding toward authoritarian rule; there will be geographical shifts in key resource availability as smaller ore and oil deposits are expended and the massive reserves (of Mid-east oil, for example) assume an ever higher fraction of the production load; and there will be changes in the priority-rating accorded by other countries—our principal allies and potential adversaries, in particular—to the needs of security and the exploitation of military power. Here the lagging trends in NATO contrasting with continued Soviet military strengthening are of key significance.

Against these stimulants of change, important forces for continuity within

39

the defense forces themselves press us to keep our interests as stable as possible, while allowing for flexibility in their force implications. The military services— Army, Navy, Air Force, and Marines—as massive, widely specialized, intricately coordinated organizations, are of course capable of change—but only at substantial effort and cost, if improved performance and effectiveness rather than confusion are to be the outcome. Shifts in interests and policies impose the need for changes in military role and structure, whether through major change of direction on the international scene, or through incremental adding of new tasks and dropping of old. Such changes translate into major program efforts: resizing, re-equipping, retraining, redisposing, and replanning the use of the force. Lead times for changes of consequence may run from months to many years, may involve up to six or eight echelons of command and direction, may affect the activities of up to hundreds of thousands of uniformed personnel, and may run to hundreds of millions or billions of dollars in costs, much of which will be spread widely through the civilian economy. Ongoing programs represent a tremendous asset and an investment in coherent management that need to be carefully husbanded.

The premium on flexibility at every level is obvious: interests so defined as to deal with the widest possible range of challenges and shifts in the security environment; policies that can serve changing emphases in our interests; and a military role and structure capable of supporting a wide range of policies. As we examine specific segments of national security interest, we must do so mindful of the need for breadth and flexibility as an aid to structuring policy and determining military posture.

For such more detailed examination, three quite distinctive sets of U.S. security interests abroad can be distinguished:

Our interests vis-à-vis the Soviet Union (and China).

Our interests vis-à-vis our major allies, specifically the NATO countries in the Atlantic area, and Japan and South Korea in the Pacific.

Our interests vis-à-vis the "world beyond"—the other nations of the world, and the world's ocean areas.

Security Interests vis-à-vis the Soviet Union

As they affect U.S. security, the dynamics of Soviet policy are deeply and strongly rooted. For the purpose of this inquiry, two deserve special mention. The first is the communist view of class struggle as the central line of historical development. The Soviets are led by it to a policy of expansion and extension of the communist system throughout more and more of the world. They see the internal divisions and difficulties of capitalism as giving rise to communist opportunities that they are duty-bound to exploit. The "correlation of forces" they see as working inevitably in their favor. In the Soviet formulation, as identified by careful students of Soviet thought and doctrine, the correlation of forces in the world comprises the aggregate of all political, social, economic, ideological, and scientific-technological factors that affect the dynamics of world politics. In the communist view, it gives an entirely different justification to their efforts to extend the communist system than to the Western efforts to resist and obstruct its extension.

The second dynamic to be noted derives from the position of the Soviet Union in Eastern Europe and Asia—regions with few natural borders, with fluid frontiers and long histories of invasions from East and West and of the rise and fall of a succession of neighboring nations (Sweden, Lithuania, Poland, Austria-Hungary, the Ottoman Empire) that dominated areas now Soviet or stood in the way of Russian expansion. The sense of potential military conflict is deeply based, and the role and strength of military forces are concerns tightly linked to national existence and communist objectives.

Ample evidence is available to give insight into Soviet interests of a security nature. As in all countries, there is a mixture of interests relating to the nation, to its governmental system, and to the regime in power; in the Soviet Union the mixture especially defies being sorted out because of the monopoly of power by the single political party and because of its attempt to identify itself with the people—specifically, the proletariat—as its vanguard.

Some of the most clearly demonstrated Soviet interests having security significance to them—and hence to outside powers—are those that serve to provide them with a secure and powerful base, one that they are fully ready and committed to protect. The authoritarian regime—a regime established by themselves, the Party, and their predecessors rather than by a voting public—exercises firm, decisive control over the armed forces at every stage from recruitment through commitment. The Soviet leaders are demonstrably prepared to protect the principle of the single-party system and its control over the Soviet people by authoritarian rule within and by violent action, if needed, abroad. The domination of neighboring countries by communist regimes subservient to, or acceptable to, the Soviets provides another important part of their security structure, as does the forward deployment of large-scale Soviet armed forces on the territories of their Eastern European allies. These deployments, so long as the countries wherein they are based remain quiet and cooperative, provide a glacis that offers important historical advantages in securing and strengthening the Soviet western frontiers. To enhance this process, they maintain a military power differential vis-à-vis their allies in Eastern Europe that overawes the latter in relation to the strength of their own forces. On their Chinese frontier, they maintain a large (forty to fifty division) ground force, air and rocket forces, and a substantial naval force in the Maritime provinces. Finally, they have achieved a partial neutralization of the nuclear threat against them, a threat that could destroy and devastate their homeland, by taking major steps to deter the occurrence of thermonuclear war—in particular, by maintaining massive nuclear missile counterforces of their own, and by negotiating restraints with the United States on actions and situations that could lead to such a war.

From this base, they are able to pursue a wide range of active and expansive outward policies. They assert leadership over the international communist movement, exerting control and influence over communist regimes and parties in other countries insofar as they are able, expanding the spread of such regimes, supporting and encouraging communist parties, and strengthening communist influences in foreign societies and foreign governments. Combining political activity and influence with military power and presence, they are committed to the support of "just wars of national liberation," and they reiterate both internally and externally that detente to them in no way implies ideological peaceful coexistence. To the contrary, it enables them to pursue the ideological struggle, i.e., to extend the span of communism to additional countries, with less concern over the possibility of Western military response. They work to prevent foreign nations from forming collective groupings and collective forces that could obstruct the outward movement and expansion of communist systems or in any way threaten the cohesion and strength of their secure base.

Just how far and at what possible costs the Soviet Union is prepared to pursue its security interests have been shown in the many confrontations with other countries—allies and opponents alike—over the years. In Cuba, they sought to buttress political influence and status with nuclear missiles aimed at the United States, but they withdrew rather than clash at sea on unequal terms with the United States, under the threat of possible nuclear escalation. In Czechoslovakia, provoked beyond their level of tolerance by Czech separatis moves and faced with a possible Czech shift to a multiparty system contrary to the first principle of Communist party monopoly of power, they finally moved with ruthless efficiency and predominance to crush the Czech regime and install a substitute. To North Vietnam they provided massive military aid, in violation of the United States-North Vietnamese agreement in the final years, but they did not intervene when the United States mined Haiphong harbor and conducted intensive bombing around Hanoi to cut off the flow of military support to the North Vietnamese.

The Soviet interests and objectives outside their own country, well supported by measures of political, economic, military, and propagandist action, cut across the security interests and needs of a great number of the other countries of the world. Though the Soviets describe the conflict (in terms favorable to themselves) as the historical process of socialist evolution and the fall of capitalism, it presents itself to the West as a well-orchestrated effort to extend Soviet hegemony—weakening, threatening, and, where advantageous, attacking all who stand in their way. They pose a threat to the West, and specifically to the United States, that is fundamental and inherent to their system. However, this threat is pressed in tangible form only to the extent it is likely to be productive of favorable results, within limits of costs and risks to which they are closely sensitive. Their doctrine, while viewing the triumph of socialism as inevitable, nevertheless recognizes that it must be carried forward by struggle and effort, and not awaited as a result to be expected to occur automatically. And they place careful limits on the struggle and effort—seeking, ideally, to gain the fruits of war without the costs or the risks of war.

From the foregoing, the broad lines of a suitable framework of U.S. security interests vis-à-vis the Soviet Union emerge with some clarity. To preserve our values, the way of life which is for us characterized by the widest possible range of personal freedom and self-government, and to keep people and economic activity safe and secure, we must deal with Soviet power and purpose where they come into conflict or potential conflict with our interests; most imperatively we must deal with the thermonuclear threat they pose to us.

Our main security interest will be one of denial. We must deny the Soviet Union any possibility of using its thermonuclear arsenal to its advantage through initiating an attack on us; we must hold this powerful and destructive Soviet force in check while assuring that our other great values are safeguarded and kept free of impairment from outside interference. The interests of denial include denial of Soviet territorial aggrandizement, of their seizure and control of free world resources, of such an extension of their influence and control into foreign lands as would carry with it serious hazard to the United States. They must be denied the achievement of a position of overall military strength and advantage from which they could, in effect, dictate terms to the United States—that is, call the tune in international affairs. We must certainly see that Western Europe is denied to them—and Japan as well. They must be denied control over Middle East oil—a particularly important task made particularly difficult for us by the weakness and fragility of the key governments of that area. Were Soviet-dominated regimes to gain control over the oil resources of that area and over the availability of that oil to the West, they would hold a life-or-death power over Western industry that could soon prove suffocating and intolerable. And there are, of course, other areas of the world as well—Brazil and North Africa, for example—whose importance to the West for raw materials or strategic reasons is such that they, too, must be denied to the Soviet domination. Many of these, we may note, are areas where the highly developed Soviet techniques of communist political organization, internal subversion and penetration, seizure of power by disciplined minorities, and imposition of authoritarian police-type controls offer them powerful means of imposing their dominance. They are areas where the restraints against Soviet (or Chinese) reach for power or territory and against Soviet (or Chinese) encouragement and support for military action are much less substantial than in the cases of Western Europe or Japan and South Korea. The Soviets have shown that they may judge the gains available to be worth the potential costs—as in Angola—especially if others, such as the Cubans, can be induced to do the fighting and pay at least the human costs. In such cases, a primary question for the United States, requiring case-by-case consideration, will be whether to attempt counteraction and, if so, whether to try to deal with the issue at its source (i.e., the Soviet Union itself), at the target area (i.e., the country concerned), or—more likely—by some mix of the two processes.

Finally, the Soviets must be denied the ability to prevent the free use of the high seas for commerce and other peaceful purposes or for free American access by sea to other free nations.

Denial of Soviet gains in areas such as those described is bound to create stress and tension between the United States and the USSR. Such tensions in turn set limits—dependent on Soviet readiness to accommodate to such denial—to the processes of detente and negotiation, which have become major features of United States-USSR relationships. So long as the Soviet system retains its basic doctrine of socialist evolution, some level of tension must be regarded as an enduring feature of the international system. The view of the Soviets that it is the capitalists who are trying to obstruct the normal course of world affairs by force is in this context no more than an euphemism for their own ambitions for domination.

In contrast to some of the other areas of U.S. security interest, this grouping of interests vis-à-vis the Soviet Union is one where the general lines of U.S. action to deal with conflicts and challenges are by now relatively well known, time tested, and generally supported by public opinion. The essentials for the maintenance of stability in the thermonuclear relationship have been exhaustively studied, debated, and argued with the USSR. Principal sources of possible instability—through technological advances in weaponry, especially nuclear weaponry, through possible political changes in Western Europe and the Mediterranean, through shifts in the control of resources in the third world—have been well identified. Even the remaining issues and uncertainties—the application of U.S.-Soviet detente and restraint to conflicts-of-interest affecting third-world countries, for example—are subjects of consultation, if not of negotiation and agreement.

Parallel American security interests relate to China, even though these are at this stage only pale shadows of the concerns that we must feel vis-à-vis the Soviets. The considerations need not be reviewed in detail since many are similar to those for the Soviet Union. The Chinese power base is, of course, much weaker, and there are undoubtedly still many years ahead before the Chinese could constitute such a threat as the Soviets have already achieved, but it is an interest and concern for the future. Whether the Chinese will in fact develop such a threat, and what tangible forms and directions it will take, are still to be learned. If the Cultural Revolution has done nothing more, it has kept fluid the matter of intensity, form, and targetting of Chinese communism in its outward aspect. But the evidence is sufficient—in Tanzania and elsewhere—to indicate that a Chinese presence is developing on a widespread basis—and that a Chinese role, at least in the world's "countryside," may grow to greater proportions in the years ahead. Already the conflict of Chinese with Soviet concepts of communist action in such areas has been clearly evidenced.

Security Interests vis-à-vis NATO, Japan, and Korea

A second set of U.S. security interests of which we must take special notice are those with its principal allies in the highly developed areas of Western Europe and Northeast Asia—Japan and South Korea.

Here the need is for the United States to work in close cooperation with allies for the attainment of shared objectives. This effort will be no less essential—and in some ways even more difficult—in the time ahead than in the past. Much of what the United States as a nation will wish to accomplish in the world can only be accomplished in cooperation with the efforts of others, and especially the efforts of like-minded nations, notably the Western-style industrial democracies, which dedicate themselves to the support of modern civilization and free political processes.

The U.S. nuclear umbrella retains its vital importance for the security of Western Europe and the Japan-Korea area, and it is a key reason for the concert of interests that exists for security purposes between them and the United States in these respective areas. Should these alliances break down, a vastly changed—and vastly more dangerous—security era would be upon us. A wholly new appraisal of security needs and possibilities for the United States in light of Soviet and Chinese attitudes and capabilities would be urgently required.

The terms of the existing treaties give concrete expression to our shared interests, and these have been augmented and implemented with further agreements and decisions defining the ways in which joint action—joint military action, if necessary—would be carried out and making joint preparatory arrangements, involving such matters as bases and deployments, on which deterrence and defense in case of attack must depend. The adopted policy is one of collective preparedness, aimed at making both deterrence and defense more effective.

The foremost American interest relating to Western Europe is obviously to deny this area, or any part of it, to the Soviet Union. Soviet possession as well as Soviet domination short of possession must both be denied. This requirement implies at once an American interest in seeing that the European nations, as many as possible, stand with their fellow Europeans to deny Soviet control over any or all of them. Not all of them do, and the absence of the neutrals from NATO, as well as the exclusion to date of Spain, necessarily weakens the security of all. Of like effect has been the curtailment by France and Greece of their participation in NATO and in the NATO-integrated military activities. In addition, the participation of communist parties in Western governments—as in Iceland and Portugal at times in the past—has limited the full flow of NATO information to such governments and has consequently curtailed their full participation in NATO affairs and their contribution to the security of all.

With respect to the individual countries, the American interest (like their own) is therefore not only to deny them to the Soviets, but to enlist them in the collective security enterprise from which all benefit.

A comparable interest exists for the United States beyond the NATO area itself, but here the sharing of interests with the European nations has obviously been much less complete, and it has varied from country to country and time to time. During the period of decolonialization after World War II, the differences of aim between the United States on the one hand and Britain, France, Belgium, the Netherlands, and Portugal at various times on the other—often sharp and stress-producing—precluded any common policy. In more recent years, notably during and after the 1973 Arab-Israeli War, the European nations showed their readiness to part company with the United States over oil policy with the Arab states and over American support for the security of Israel. America's NATO Allies, the Scandinavian countries especially, have characteristically been extremely reserved about concerted action concerning problems outside the NATO area (or even consulting on it) within the NATO context. Undoubtedly, their reluctance reflects in part the desire to avoid responsibilities beyond those that are purely domestic or European; but, in part, it reflects differences between themselves and the United States regarding the policy stance or image they wish to maintain with third-world countries—in particular, their desire not to be saddled with some of the decidedly negative attitudes toward the United States that are found in those countries and especially among their political activists.

In any case, experience has shown over a lengthy and varied period of third-world development that the United States can expect from its NATO partners little help or association—but often a significant amount of opposition—in trying to give assistance to countries such as Israel, South Vietnam, and Angola against external, communist-supported threat and attack.

Within Europe itself, just as in the case of U.S. interests relating to the Soviet Union, the main lines on which solutions to the problems of the alliance can be found are well known and widely accepted. There are persistent stresses and challenges, however, inherent and inescapable in the fact that the alliances represent a sharing of important values and interests—substantial, but nevertheless far from complete—and not an identity of such interests or values, nor anything like a uniformity of national traditions. Neither is a consensus complete or permanent within any country regarding such questions as the benefits and costs of the alliance, the suitability of its forces, its basing, training, or other activities, or the priority that these activities should receive in relation to other budgetary needs and social pressures, e.g., for the reduction of conscription. Acute differences such as the quarrel of Greeks and Turks over

Cyprus and the Aegean sea-bed or the British and Icelanders over fishing areas likewise impair essential solidarity and give openings for Soviet initiatives to weaken the alliance. Problems and challenges outside the recognized treaty area, or outside the customary scope of matters for joint action, pose a particular problem of policy coordination highlighted, for example, during the oil embargo of 1973, the Arab-Israeli war, and the resistance to proposals to extend naval surveillance to cover the oil supply routes around Africa and through the South Atlantic.

Important unresolved issues remain within NATO, reflecting real or perceived divergence in the interests of individual nations, e.g., in the failure of NATO to achieve rational standardization of arms within the military forces of the member countries. Nevertheless, and despite these differences and remaining issues, the area of shared interest is large, and it is well understood and supported by a broad consensus in every country, including notably the United States.

Security Interests vis-à-vis the World Beyond

In sharp contrast, U.S. security interests vis-à-vis other nations of the world—that is, those other than the USSR, China, our NATO Allies, Japan, and South Korea—are nowhere nearly so well defined, well understood, or generally accepted. The reasons are many. First, the countries themselves are more numerous—a hundred or more—and are widely varied in size, economic status, degree of self-government, social system, attitudes toward the United States (and the USSR), vulnerability to subversion or other forms of outside threat to their sovereignty, etc. Some have special ties to the United States; others have none. Some have vast stores of oil and other critical, valuable raw materials; others have little or none. Some are at an advanced stage of development; others have barely begun. Some, such as the Latin American, have many close contacts with the United States; others are remote, and have few or none. There is a fluidity about policies, alignments, and national goals that has thus far defied stable, enduring relationships. It is far from clear what degree of order can be maintained and safeguarded in such a heterogeneous and shifting population. U.S. policy itself has been uncertain, inconsistent, unsettled by massive failures as in Vietnam, and blocked by unreconciled disputes between Congress and Executive.

From President Truman through all his successors to date, no administration has found a policy giving assurance that it would be at the same time adequate and supportable. The Truman doctrine supporting freedom wherever attacked in the world, the Kennedy pledge and policy of ready and flexible response, the Johnson commitment in Vietnam, all have proved in retrospect to exceed what our country was likely to support. The Eisenhower approach, limiting action more sharply to a few selected crisis points—Lebanon and the Middle East, Taiwan and the Formosa Strait—together with the Nixon doctrine enunciated at Guam, undeniably left open the possibility that support thus limited would not prevent the fall of free nations, as the fall of South Vietnam and subsequently the takeover of Angola have since demonstrated. This is the dilemma that each postwar president has had to confront. It is a dilemma that still exists and seems unlikely to be overcome, since hard experience has now shown that what may be required for nations to survive may exceed what they—and the United States— would be able and willing to provide.

In fact, it is a significant source of international and security concern that there may be a widening gap between the need for world order, particularly among the weaker, poorer, and smaller nations, and the means available for fulfillment of the needs. The danger is that they will be left adrift, without possibility for economic progress and national safety. The United States may not have been the "world's policeman" but its withdrawal from the role of partial support it had assumed seems likely to mean greater vulnerability and instability throughout the free world.

The direct significance to U.S. security in a military sense of individual countries in these areas is correctly seen as substantially below that of Western Europe and Japan. Nevertheless, the U.S. trade with many of them has major implications for our security, notably with the Middle East oil-producing states, and with others producing other essential minerals and raw materials. Our concern for access to bases and for safe passage through narrow waters adds to the importance of others. The broad worldwide U.S. interest in the principle of sovereignty of free nations creates another basis of friendly ties. Special ties with Israel, the Philippines, Australia and New Zealand, and our Latin American neighbors add to the list of interests. In other particular nations, past U.S. commitments to economic progress and to advances in health and education provide further source of possible U.S. security interest, as do military aid and cooperation provided in support of recognized shared interests.

There are two principal ways in which events and conditions in countries of the third world can impact more heavily, and adversely, on U.S. interests. The first would be the coalescence of several countries into a bloc or group powerful and hostile to the United States. By population, resources or strategic position, countries such as India, Indonesia, Nigeria, or Argentina—particularly if joined by several neighbors—could impose significant handicaps or deny significant opportunities to the United States. In general, however, so long as these actions were taken peacefully, they could be accommodated or gradually ameliorated through normal efforts to improve relations, providing they remained free of Soviet (or Chinese) connection.

It is the bringing in of Soviet (or Chinese) involvement that would give sharply added significance to such a development. Short of this, it is rare indeed—oil being the sole apparent exception with any appreciable probability—

that one can conceive of the possibility, after the Vietnam test of the limits of U.S. interest, that armed U.S. intervention would be initiated unless our forces or our shipping were militarily directly attacked. This is not to say relations with the countries of the world other than major powers and allies should or will be neglected. Rather it is to recognize that contacts are likely to have less of security factors and issues in them, particularly those of a military nature, and more of the economic and political—again, so long as there is no significant Soviet (or Chinese) connection and control.

Typically then it is the three-way relationship of such countries with the USSR (or China) and the United States that raises their security interest to the United States. We will be concerned for their health and viability as sovereign states that should remain so. But it is where the danger exists of their coming under Soviet control, of their contributing forward bases and resources to the deployment, presence, and operations of Soviet naval and air forces (especially those in position to dominate or obstruct shipping routes), or of their interfering with U.S. deployments and freedom of movement, that closer calculations of gains in Soviet strength and reductions in our own become necessary. And this is precisely the area—well demonstrated in the Angola experience in 1975 and 1976—where the lack of agreed American policy leaves the United States without a working yardstick for determining the threshold at which U.S. interest would translate into U.S. action.

The broad U.S. interest is to prevent countries of the third world from being forced by Soviet pressure into hostility to us, or into providing an added increment of strength to the Soviets. This is by no means an area, however, in which the United States can, in general, dictate or dominate. Rather the approach must be one of negotiation, inducement, and encouragement of friendly and positive attitudes, since we share with them—as the most important and fundamental interest—a desire to see their sovereignty maintained. If they, for their own reasons and at their own initiative, nevertheless wish to take a posture hostile to us, that is their undoubted right. We must respect this action so long as it remains within peaceful limits—even while we work to persuade them of the error of their ways, and the dangers of "supping with the devil."

One tends today to look in vain—whether to government, to centers of learning, or to the substantive facts themselves—for a full or convincing doctrine and policy by which the United States can confidently conduct its dealings with nations of the third world insofar as security problems and interests are concerned. Yet even if completeness and full assurance are not attainable, it does not seem impossible to lay down some guidelines by which we can maximize our chances for favorable outcomes in the cases most important to us.

The nations we are discussing have, as noted, all too little in common. Any notion of a single policy for uniform application to all thus seems certain to be illusory. Something simpler, more closely confined to the fundamental issues, seems clearly necessary.

What the nations as nations *do* have in common, for all their diversity, is their independence. Their interest in preserving it is an interest the United States shares. Respect for their sovereignty—their freedom from outside interference or coercion—can be the best basis for a policy that distinguishes the United States from the Soviet Union, particularly insofar as self-government exists in the country concerned. While a ruling regime might accept a compromise of sovereignty for the traditional "mess of pottage," or to keep themselves in power through outside assistance, an electorate with substantial power and voice seems less likely to do so. In many of the newly independent countries, still vulnerable to the possibilities of Soviet penetration, the sense of nationalism and sovereignty provides good grounds for hope that they can survive.

If sovereignty is taken as the basis for our policies and interests and the proper road for development of our basic relations, several other false roads will need to be closed off.

Cold war is a term used to describe the opposition of U.S. and Soviet interests, both directly between each other and as they apply to other countries. It is a biased and inadequate description. If U.S. interests center on the continued sovereignty of the nation concerned, the latter is by no means a mere pawn in a great power struggle. The crucial element to its success—indispensable, even though possibly insufficient within its own resources, if attacked from outside—is its own will and determination. The United States, even within the constraints of the effort our public will support, can at least stand with the nation in principle. The stance of Soviet communism is quite different—there is in it an implication of hegemony, the reduction and subversion of the nation's sovereign independence.

Likewise, ideological difference should not be allowed to characterize the U.S. relationship with such a country—specifically, to drive it and ourselves into an adversary posture. So long as its sovereignty is preserved in ways that avoid its becoming a tool of the USSR, we, at least, would seem to have no adequate reason for regarding it as an opponent. We have to recognize that nationalism often needs something to be "against," and American economic power makes it a natural target for charges of economic exploitation, especially from leaders who tend to be sympathetic to Marxist theory. Nevertheless, it should be possible to make known our views that state-run economies are inferior to systems of free enterprise, and that representative government is preferable to authoritarian rule, without driving authoritarian countries with state-run economies into a hostile camp. We must expect a certain amount of verbal attack and political abrasion from nations so structured—in particular, from the leaders who benefit from the operation of their system—but so long as they do not subject themselves to Soviet domination, our most important security interest seems likely to be generally met.

Even while the drift toward authoritarian, nondemocratic rule goes on, however, the American example and leadership in self-government retain an important weight. American moral leadership is considered as sustaining democracy and the principle of self-rule in the world. (One result of this is a special risk for the United States in associating with totalitarian regimes.) By its moral stance the United States can dissipate some of the uneasiness and distrust that is inevitably generated by our sheer political and economic weight.

The effort to represent U.S. foreign aid as attempted U.S. domination has been a Soviet theme since their rejection of the Marshall Plan and their insistence that Poland and Czechoslovakia also drop out. The proper test for the recipient country to apply ought to be whether the aid, on the terms offered, is or is not within its sovereign interest—not whether it is within the U.S. interest. A valid reason for the aid program has been that it serves the interests of both giver and receiver. As a corollary for the United States, it is important to avoid any action violative of the recipient's sovereignty, since that is the factor that differentiates U.S. aid from the kind of Soviet aid that causes concern (a concern, it should be noted, not solely of the United States, but on occasion, of such significant recipients as Egypt, Syria, and Libya). A problem previously mentioned is again encountered here. Over the years, a subtle shift has often occurred in which recipient nations go beyond welcoming the support and cooperation of the United States to taking it for granted, demanding it, or even trading on it by seeking higher bids from the USSR. Also, foreign aid benefits not only the nation receiving it but often its leaders and particular political groups as well. Politicians everywhere find it natural and beneficial to confuse their own interests with the nation's interest. The likelihood of dissatisfaction and argument on their part regarding the terms of U.S. aid must be recognized. It becomes more serious if groups form in opposition to the U.S. over differences in the amounts, the terms or the channels and processes of aid. It will remain a U.S. interest to restrain the formation of such groups, since with such organization they could readily shift to actions—such as the cutting-off of raw materials, or interference with ocean routes—that could be damaging to the United States.

Whereas these and other U.S. security interests in these areas are best pursued by nonmilitary means—and by processes indigenous to the countries concerned—there remains always the possibility that stronger measures may become necessary. Obviously, such possibilities—contingencies rather than probabilities—impose complications for U.S. military role and posture, to which we shall have to turn at a later point.

Here, however, we take note of the need for development of mutually advantageous relations with all nations of the third world—allies, neutrals, and even adversaries—relations that will bring benefit to all sides. Otherwise we will find ourselves increasingly faced with a rising tide of animosity and interference on the part of the underdeveloped, poverty-stricken nations of the Southern Hemisphere—nations that certainly can be mobilized in ways damaging to world order and peaceful trade. A high degree of freedom of the seas will remain essential for protection of the commerce on which modern industry depends.

The elements of U.S. security interests that should be recognized therefore would seem to include:

support (within practical limits) of national freedom, and of strengthened indigenous nationalism;

safeguarding the free use of the seas;

striving to maintain the essential minimums of order and stability incident to expanding trade, commerce and economic well-being, in an era of increasing economic interdependence.

With all of this, it must be recognized that distrust, even antagonism and hostility, will undoubtedly be evidenced toward the United States by the governing regimes of many countries. So long as it does not break out into violence against us, or result in the forced entry of such countries into the "socialist camp," it must be recognized as part of the world environment not subject to our control. It must be seen also as neither critically threatening nor harmful to the security and safety of our people.

4

Policy Paths to Security—
Defense, Deterrence, Detente

There are three principal routes—distinguishable as defense, deterrence, and detente—by which to make provision for our national security in its military aspects. Each constitutes an important line of policy by which to pursue and provide for our security interests. Military power—the strength, composition, and specific weaponry of our forces and their posture of readiness—contributes and responds to each of these lines of policy.

Defense as here used refers both to actual military conflict and to the military preparations for such conflict—the employment of armed forces in combat operations, whether at the level of all-out war, or for limited intervention or local self-defense. As a policy it covers both the fighting that would take place in event hostilities of whatever kind should occur and the military establishment that we build and maintain to meet that fundamental, ultimate need.

Deterrence acts (in a considerably more complex way) to persuade a potential enemy—because of costs, risks, lack of gain, and uncertainty of outcome—not to use force against us or our allies, nor even to threaten its use for purposes of interference or coercion. More broadly, we may seek to deter lines of action—of whatever kind—that would jeopardize our security interests, and may even exert pressure on a potential enemy to cease and reverse actions, such as the Soviet introduction of nuclear missiles into Cuba in 1962, that were already well underway before we began to react.

Detente, in the sense used herein, is aimed at reducing tension between opposing states (the United States and the USSR—or China—in particular), thereby ameliorating or normalizing the relations between them, containing or reducing the threats each projects to the other, lessening the chances of being drawn into mutual conflict by third-area issues, and even (as envisaged by some of its proponents) building up a body of mutually beneficial commitments and exchanges—political, economic, technological, cultural, and the like—that each would be reluctant to see destroyed.

The contributions that are made to security by defense, deterrence, and detente take distinctly different forms. In dealing with a potential opponent, for example, defense is designed to meet an opponent's attack with overt military action, while deterrence is intended to meet an opponent's threat in ways that forestall an actual attack, and detente is aimed at reducing an opponent's threat through, for example, mutual force reductions and the resolution of mutual differences and disputes.

Similarly, the significance of the three policies in terms of relations with our allies is quite different. If we think of actual defense, specific local military vulnerabilities weigh heavily among our concerns, as do local force contributions and collective arrangements to maximize military effectiveness. For deterrence, added importance is attached to *allied solidarity*, which emphasizes to a potential opponent that an attack on one is likely to be an attack on all; more reliance tends to be placed on nuclear arsenals by virtue of the dissuasive power in their great destructive potential. Detente in turn is two-edged: its contributions to reducing the threat will be welcomed by our allies; but the risks that agreements between the United States and USSR may weaken the effect of the nuclear deterrent, or endanger exposed territorial areas through reduction of U.S. presence and reinforcement capability, will be uneasily watched and calculated.

In the third areas of the world where a U.S. policy of defense—the actual use of military forces—would be synonymous with big-power intervention and likely to result in large-scale destruction and loss of life, there will be a strong reason for the countries of the area to hope for the prior success of efforts at deterrence (specifically, the prevention by peaceful means of Soviet encroachment or seizure of power). As a policy for the United States, detente seems to offer small counterweight to Soviet support to "wars of liberation," i.e., takeover by force. It carries as well a lurking implication of "spheres of interest," with small countries seen by the largest powers to be the objects of each other's policies, rather than independent actors in their own right.

No one of these policies—defense, deterrence, or detente—can today by itself offer to the United States (or to the other nations involved) an adequate or economical means of providing for security. Even if American defense dispositions were the densest and most massive that could possibly be imagined, they would still be so porous to Soviet air, submarine, and missile attack as to leave millions of our people and vast portions of our national wealth vulnerable to attack and destruction. Deterrence in turn depends crucially on a real capability for defense, and detente is inherently limited in the reliance that can be placed upon it by the constant possibility that one participant or the other may decide that a challenged interest outweighs the risk of harmony; so long as each disposes of powerful military capabilities, in the form of forces readily employable with little warning but great destructive effect, the ability of each country to shift its policy and modify its intentions with little notice sets sharp boundaries on the application of this policy.

But while none is sufficient, all are necessary to an effective and sensible security posture. They are not, of course, entirely or perfectly compatible—a weapon such as the antiballistic missile that improves defense may have destabilizing effects that could add to tensions and work against detente. But to a large extent each can in fact support the others. Our security arrangements, including the design of our armed forces, must reflect a correct understanding and weighing of their respective roles and help to make the support that each gives the others as effective and as dependable as possible.

Defense

To be prepared for war in time of peace has been a central strand of security policy since World War II. The outlays for such preparations have been immense—something over a thousand billion dollars for U.S. defense budgets alone from 1945 to 1975, at least an equal amount for the Soviet Union if expressed in comparable terms, and probably no less, in the aggregate, for the remaining nations of the world. If a single figure were to be placed on the world's total outlays for a single year today, it would have to be (in the same terms) in the order of $300 billion, or even more.

Several hundreds of these billions have been spent in actual hostilities—the communist insurrection in Greece, the Korean and Vietnamese wars, the Soviet intervention in Hungary, the Indo-Pakistan and Arab-Israeli conflicts, and numerous other lesser clashes. But the vast bulk of the aggregate trillion-dollar outlay has gone to support the warmaking function by being prepared for war, not by fighting it.

Nevertheless, because the other policies—deterrence and detente—are always subject to failure, it remains necessary to design and support forces adequate to fight war effectively if it should be thrust upon us. The basic task of the defense strand of our policy is to "protect our national valuables" (in the apt phrase of General Maxwell Taylor) in case of war—by the effective conduct of military operations, paying their costs, bearing their burdens. We would meet enemy attack with our own military response.

In the United States, the main structure of our defense forces emerged soon after Korea, being principally enlarged and augmented by the addition of the ICBM and Polaris forces in the late 1950s and thereafter. Since the Eisenhower Administration's "New Look" in 1953, the nuclear weapon has played a dominant role in defense planning for the United States, and this dominance seems unlikely to change. A second compelling factor, almost equally powerful in shaping U.S. defense preparations, has been the perceived need for *collective* defense arrangements—the association in various forms of the United States with other independent noncommunist nations. Here the accent has fallen heavily on conventional forces, though not exclusively, since tactical nuclear weapons also have an important and essential supplementary role to fill (in NATO, for example). Except for the ultimate possibility of mutual direct nuclear attacks between the United States and the USSR, defense against major attack in the West or involving the Western powers has meant, in some sense, collective defense. This need exists in strictly war-fighting (as well as deterrent) terms; it has been reflected in the spawning of arrangements for collective defense—

arrangements that extend from NATO, with its organized institituions and its force dispositions in place, through other multilateral and bilateral undertakings, such as CENTO, SEATO, and the Japanes-U.S. Treaty, to the more limited agreements for mutual assistance and military aid with forty or more countries.

On the Soviet side, despite occasional fluctuations of limited scope, a large total military establishment has been maintained at all times without exception since the war. Efforts have been concentrated on ground forces, air defense, rocket and missile forces, an expanding Navy, and more recently on the buildup of the large forces now deployed near the Chinese borders in Asia. A carefully elaborated military doctrine has been prepared, which guides the development of forces and lays down the major lines of operations that the Soviet armed forces (supplemented by their allies) should be prepared to perform in case of war. The Soviet writings on this score have been extensive and provide clear guidelines for war. It is of interest, however, that the doctrine essentially applies *after war has begun*, i.e., the discussion tends to omit consideration of the "crisis management" stage of potential or threatened conflict—a significant omission indeed, since this would surely be a crucial period for security policy in a time like the present when actual resort to force by the United States and the USSR against each other could be nothing but a choice between extreme evils. Likewise, the doctrine gives little consideration to the probable linkage between a massive, surprise use of Soviet nuclear weapons in the event of war in Europe (which is what the doctrine contemplates) and a nuclear exchange escalated to the strategic level that would involve the Soviet homeland itself. Thus, while the doctrine has obvious utility for the purpose intended—force design, force training, and force readiness, in particular—we may note that it leaves open several tremendously important questions: how and whether war would begin and whether the prescribed launching of the nuclear assault in Europe would in fact be carried out, once the dangers of escalation to all-out exchange with the United States were taken into account.

We on our side have built and hold at varying stages of readiness a multicomponent military force that is designed to provide a powerful counter to the Soviet military power.

The overriding task is to counter Soviet nuclear attack. To prepare for such a contingency, we have required our own forces to be able to conduct nuclear retaliation as an assured certainty, or nuclear preemption as a less likely possibility—a possibility, for example, in case we were to receive unmistakable warning such as a swarm of hundreds of Soviet ICBMs actually launched and in flight against our territory. More recently, a third requirement has been imposed: to be able to carry out a series of more limited attacks with strategic nuclear weapons as part of a possible series of exchanges by the two sides against selected military targets, rather than urban/industrial objectives.

As part of our preparation for actual war-fighting, our nuclear force, including Minutemen in hardened silos and Polaris submarines at sea, is charged with giving us a reasonable expectation, in case of all-out nuclear exchange, of being able to terminate it in a position not inferior to that of the Soviets. In addition, the strategic nuclear forces must provide us a capability of conducting "counterforce" attacks against Soviet nuclear forces and other military targets (as they can against us). The number, disposition, accuracy, and multiwarhead capability of our missiles give reasonable assurance, for the present, that we would not be doing so at a relative disadvantage. A major effect of such counterforce attacks by our strategic forces would be "damage-limiting," i.e., we would be able to reduce—to a degree, but only to a degree—the damage that the Soviets could inflict upon our urban/industrial complexes. But it must be understood in this regard that no "counterforce" capability that is imaginable today could conceivably reach "first-strike" proportions, i.e., knock out so much of the Russian ICBM and SLBM arsenal as to leave them unable to respond. Nor is such a capability available to the Soviets. It must be acknowledged further that strategic forces with the limited capabilities just described, awesome though their destructive powers certainly are, would nevertheless fall far short, if such conflict should actually occur, of providing the kind of safety for our people that we would like to provide. The shortfall—the remaining vulnerability—is the inherent and inescapable burden that has been placed on us all by the irreversible development of modern nuclear arsenals.

The second major requirement under the policy of defense is to defend areas such as the North Atlantic/Western European region and the Western Pacific/Northeast Asian area (Japan and South Korea, in particular) against hostile attack. The policy requires first that we prepare ourselves to hold as large an area as possible against every form of aggression; but beyond this, our aim must also be to do so with the kinds of military weapons and tactics that will bring hostilities to as early an end as possible, with the lowest possible loss of life and the lowest possible level of physical destruction. In these crowded, industrially developed areas, with hundreds of cities and cultural and social treasures, this is a forbidding task, yet an indispensable one. It cannot be evaded, since it is the way we and our allies would protect the territorial integrity of these vital regions—our national homelands—and it provides the very basis for the policies of deterrence and detente we will be discussing later. It is an essential contribution to the national security interest of denial, which we earlier placed in the highest order of priority—denial of Soviet territorial acquisition and of Soviet political dominance over these vital centers of the West.

Nevertheless, it must be clearly acknowledged, in appraising the policy of defense—i.e., the actual fighting of the war—that its limits are severe indeed in terms of fulfillment of security interests. If large-scale military conflict were to occur in Western Europe or in Japan and Korea, the clashing conventional forces would penetrate deeply on the ground, in the air, and at sea. They unavoidably would cause heavy devastation and civilian loss of life—in addition to tremendous destruction of military forces and equipment on both sides. If nuclear operations were added, then further extensive destruction and loss of life would undoubtedly be suffered once they went beyond very limited and localized attacks. It is, however, conceivable—although not, of course, by any means assured—that the initial limited use of tactical nuclear weapons could be so telling in its effects on the enemy as to bring a halt to attacks, ·lead to a cease-fire, and thus cut short the losses that continued conventional conflict would otherwise certainly inflict. A final point is that the conventional forces alone do not give assurance of preventing the overrunning of Western Europe, nor of South Korea as a preliminary to an attack on Japan—an assessment well known both to the West and to the Warsaw Pact nations. It is always possible that the Western combat operations would have to be augmented by nuclear fire-power. Even then, there can be no certainty that the enemy's ground attack could be stopped, although the odds of doing so would certainly be increased.

The third potential application of a policy of defense—preparation for and conduct if necessary of actual combat operations—is in the world beyond. Intervention operations, possibly coupled with counterintervention operations against Soviet or Chinese initiatives, remain at least a theoretical possibility. And sea control—action to safeguard the free and secure use of the high seas and to protect peaceful shipping along the major shipping lanes—is a further implication of the policy of defense. We know that the inhibitions after Vietnam are strong indeed against American intervention or commitment in distant areas, particularly for indistinct interests and with uncertain prospects. Nevertheless, it would not be enough simply to say "No more Vietnams," and drop all requirement for military capabilities suitable for intervention operations from our defense policy. These are turbulent areas of the world—some of them, like Latin America, close to the United States—where events can impact swiftly and deeply on American interests; a new generation of American opinion could quickly arise that would support and demand a more active response to emerging problems.

Summing up the essentials of the policy of defense, we recognize the need for a strong, ready military force and for the skill and the will to use it if necessary. The measure of the forces needed can best be determined through realistic and detailed military planning by responsible, experienced military commanders and their staffs and reviewed by informed analysts. A well-grounded military writer—Drew Middleton of the *New York Times*—has asked a key question in his book *Can America Win the Next War*? His analysis leads to a gloomy conclusion, although it points the way to practical remedial actions that could restore a better balance. As a great power, the United States is obligated to maintain the confidence that it can defend itself and join in the defense of its allies, if and when attacked. Perhaps equally important—and this is certainly where our principal hopes and preferences must lie—the capabilities for effective defense and for waging war as successfully as possible are at the same time indispensable underpinnings for each of the other two major lines of security policy: deterrence and detente.

Deterrence

Since World War II, the somber estimate of the limits on what could be done toward assuring security by defense alone—in the sense of the actual waging of war—has led to consideration of what deterrence, underpinned by a substantial defense capability, would be able to achieve for us. Here the picture becomes much brighter. Our efforts at deterrence have in fact provided remarkably good returns in terms of enhanced security for the noncommunist nations, large and small, that have been faced with Soviet power and expansionist tendencies.

The concept of deterrence is a newer one, however, than the historically more familiar resort to conflict, and it is not so generally understood, particularly in its application to specific security interests and challenges, and to the shaping and sizing of military forces.

Deterrence since the war has been effective, so far as the major nations of the West are concerned, against all forms of feared Soviet attack: the classical all-out Soviet assault across Western Europe, the seizure of Hamburg or engulfment of West Berlin, and the threat that "rockets will fly" voiced by the Soviets during the Suez crisis in 1956. Where deterrence has failed—for example, in 1950 in preventing the attack of South Korea by North Korean forces—earlier U.S. official ambiguity about our "defense perimeter," not the policy of deterrence itself, is often cited as the determining factor. Within the NATO area, there has in nearly thirty years been no failure of deterrence. Soviet military action against NATO forces and territories has been deterred. Neither has there been any effective Soviet use of threats to employ its military force, for purposes of intimidation, interference, or coercion. The two possible exceptions that come to mind—Soviet pressures and threats over Berlin in 1958-1959 and during the Suez crisis two years earlier—fell far short of compelling NATO compliance with their demands. The Berlin threat remained implicit and ambiguous, insofar as significant military action was concerned, and the quality of ultimatum contained in the Soviet pronouncements at the time was removed following Khrushchev's meeting with Eisenhower at Camp David. Most of the steam went out of the Soviet drive with the construction of the Berlin wall in August 1961, substantially ending the flight of East Germans and East Berliners to the West, which had evoked the Soviet initiative. During the Suez crisis, although there were French and British expressions of concern over the Soviet threats, the Soviet statements were seen within the U.S. government as no more than appeals for favor in the Arab community and elsewhere. The expressions of allied concern were at least partly intended to put pressure on the United States. In any case, a strong public rejoinder was made by the United States, and the issue was shortly made moot by the early UN cease-fire.

There were many other conflicts of course that were not deterred—the Soviet interventions in Hungary and Czechoslovakia, the North Vietnamese attack against South Vietnam, and the Arab-Israeli and Indo-Pakistan clashes, among others. However, none of these involved the Soviet Union in direct hostilities against the major Western powers.

In analyzing the future role of deterrence we may note that there has been general consensus—in NATO and elsewhere—that the deterrence achieved in the past has been based largely on the actual war-fighting capability of the West and on the costs, risks, and uncertainties it presents to an aggressor; no substitute for this capability can be found. But there exists in addition, particularly in Europe, a somewhat different set of ideas—a dissenting view to the effect that deterrence of the Soviets could be better achieved simply by establishing a firm and unbroken escalatory linkage between the onset of any large-scale military conflict in Europe, as a result of Soviet aggression, and the launching by the United States of its all-out strategic nuclear attack. One important factor behind this difference is a continuing European uneasiness over the possibility of a U.S. shift toward isolationism or unilateralism and a desire to bind in the U.S. nuclear forces more tightly.

This is an important divergence that has underlain a number of U.S.-French strategic disagreements. It is noteworthy that the ideas have not proved compelling nor generally convincing thus far, and they seem unlikely to do so, principally because the premise is highly hypothetical and contrary to several important realities. Escalatory linkage to be firm and unbroken would have to be automatic, and so far as U.S. weapons are concerned, no U.S. president is likely to be committed in advance with such rigidity to the release of such an attack. To the contrary, the obligation under the law to decide on such release would appear to require a president to consider the specific facts of specific situations at the time. For presidents to fail to utilize every conceivable means that they and their chief advisors could envisage or devise to terminate conflict on acceptable terms at levels of destruction lower than all-out nuclear attack is beyond reasonable expectation. Even after conflict had begun, they would unquestionably weigh the use of conventional forces and tactical nuclear weapons as possible alternatives to the strategic forces. For these reasons, grounds are lacking for the assumptions that some have voiced to the effect that a weakening of conventional forces or of tactical nuclear strength would somehow add to deterrence. In an era of nuclear parity, if we were to show a possible aggressor that we had nothing but strategic weapons with which to respond to attack, the aggressor would have good reason to see this posture as inhibiting us from any effective action at all.

The deterrence concept, despite its undoubted strength, is, however, subject to one serious shortcoming that must be clearly noted. While the *sufficiency* of our military strength to provide deterrence can be shown by the continued absence of war and coercive threats, no similar test is available to show that all of the military strength we maintain is *necessary* for the purpose. In the absence of such an overall test, there is no substitute for analysis and debate taking up each major component of force in turn for critical examination and challenge—

strategic nuclear forces, land based and sea based; theater-type forces, land, sea, and air; naval forces for broad sea-control missions; and all the supporting paraphernalia of bases, depots, repair facilities, training and logistics systems, communications headquarters, intelligence activities, research and development functions, and the like that modern forces require.

The debate cannot be avoided. Defense costs that have reached more than $100 billion a year in the United States alone leave no choice. And even at such levels, the budget squeeze—from rising real costs of more advanced weapons, rising wages, and continuing inflation—forces hard choices between alternative weapons systems, between nuclear and conventional forces, and between concepts of forward defense and massive retaliation. The only convincing proof of necessity is to show that, without such forces, our capability for defending ourselves will be left inadequate or endangered and that the Soviets (or other potential adversaries) are likely to perceive our weakness and act to exploit it.

In these circumstances it is hard to obtain the sustained, steadfast support needed to keep the military forces modern and strong. Perhaps it is in fact more noteworthy, in NATO for example, that the force structure has been maintained over the years than that efforts are made from time to time to make unilateral force cuts or that a kind of nibbling process has eroded terms of service, readiness standards, and levels of supply. The danger to deterrence is, of course, that from dispute and lack of consensus over what military forces are *necessary*, the governments will fail to provide what is *sufficient* to maintain stability. The strategic nuclear balance and the balance in NATO are where the stakes and the risks are the highest.

This difficulty is inherent in the policy of deterrence, which nevertheless remains an indispensable component of Western security, and it will remain so as long as the East-West confrontation continues. The difficulty puts a special premium on rational preparation of the defense program and on its lucid and persuasive public presentation. But it imposes, at the same time, requirements of sober and constructive responsibility on governmental leadership (both executive and legislative), on opinion-forming institutions (both academic and journalistic), and on the citizenry at large.

While there is and can be no assurance that such a sense of responsibility will always be forthcoming in the measure needed, the very success of the deterrence policy in the past at moments of crisis and peril shows that attitudes of this kind are not lacking. There can be no doubt that in the crisis over Berlin from 1958 through 1961, the collectively organized Western military power available for a broader response—had the communists attempted to take the city by force—exerted a telling dissuasive restraint on the Soviets. At the time of the Cuban missile crisis in 1962, the Soviet decision to withdraw was patently a response to the superior military capabilities and the demonstrated intent of the United States.

Such capability and intent on our side are two of the three principal essentials for the success of our deterrent policy. The third, on the other side, is Soviet belief that our capability and intent are such as would impose on them (in case hostilities should occur) costs, risks, and uncertainties they would find unacceptable in relation to the gain they would be likely to achieve. In cases wherein the Soviets have not been deterred from actions contrary to U.S. interests or commitments—in supporting North Korea or North Vietnam, for example—they have obviously believed that the United States would not intervene or that, if it did, its intervention would not escalate or extend to a point that would involve direct U.S.-USSR conflict. Although they were wrong about U.S. intervention, they were correct in thinking that it would not escalate to involve them in direct conflict.

A further, final word may perhaps be said on the basic concept of deterrence. In an article prepared for the magazine *Orbis* in Spring 1969—in the issue commemorating NATO's twenty years—a Soviet contributor, Sergo Mikoyan, observed that there had been no day in those twenty years in which NATO could have defended itself successfully. He observed that it must therefore have been something else that safeguarded the peace in Europe. His comments show how important it is to apply the right tests and the right criteria when evaluating deterrence. The proper question was not whether NATO could have defended successfully; it was whether the USSR could have attacked successfully. Despite all the shortfalls and weaknesses in NATO's posture, one need not doubt that the Soviets saw costs, risks, and uncertainties that made any notion of attack unappealing. We may note, however, how different was their action in Hungary (1956) and Czechoslovakia (1968) when the disorders (and political heresies) that arose confronted them with potential threats of a higher, more immediate order. These were events that potentially threatened the continued solidarity of the Eastern European military buffer from which Russia so greatly benefits as well as the continuance in power of regimes subservient to them (which threatened to establish precedents of multiparty systems that could conceivably or ultimately infect Russia itself). At these times they did not shrink from the decision to take military action, and when they did so they launched massive preponderant forces with little if any visible concern over possible military response by the West.

It is obviously not given to us to know precisely how the Soviet Union will calculate the deterrence equation or what their specific conclusions will be in specific instances. Too much of the decision process is subjective. We can, however, identify the principal factors that will necessarily enter into it and form our own judgments regarding their influence under a relevant range of scenarios.

First is the prospect of gain. We know enough of the dynamics of Soviet policy to accept that they place high value on the extension of Moscow-guided communist control over more and more of the peoples of the world. Undoubtedly, this motivation was at its strongest when it was applied to Eastern Europe just after the war, for there it was combined with the possibility of establishing friendly regimes on the territories from which the great attacks against the Soviet Union had been launched in the past. Another relevant version of gain for them is prevention of loss. Much of Soviet postwar conduct shows how highly they value what they have achieved in Eastern Europe. It is there, above all, that they have applied the Brezhnev doctrine to discourage any hopes that nations once brought into the "socialist camp" will be allowed peacefully to slip away.

Against any potential for gain that they see must be arrayed the prospects of cost and risk. Modern war is staggeringly expensive. The intensity of the Middle East war of October 1973 has shown what the scale and rate of immediate battlefield losses can be, and they know that vastly larger forces would be involved in any full-scale clash with the West—close to a million on the communist side alone, for example, in just the initial phases of conflict in Central Europe (East Germany, Czechoslovakia and Poland). And beyond the soldiers would be the whole civilian population. The costs of conflict would come quickly, and they would be great. The experience of World War II with so many millions of soldiers and civilians lost in conflict gives the dimensions—twenty million or more—of what large-scale war in Europe could mean. And thermonuclear war—a conflict in which each side launched its inventory of strategic weapons against the other—could cost at the least very close to a hundred million lives in the Soviet Union alone and might go considerably higher. The devastation and destruction of industry and cities, not to mention the lasting environmental damage, would be of comparable magnitude. The spectrum of possible costs is wide, and the risks of escalation from even a limited initial clash are severe.

Uncertainties impact heavily on all the factors that an aggressor would have to evaluate—the gains, the costs, and the risks. Because the stakes are so high, they powerfully reinforce the deterrent. Modern war is so complex, and the outcome so sensitive to even small actions at crucial times—the breaking of the Japanese code at Midway, the Israeli crossing of the Suez in the October 1973 war—as to impose a heavy additional burden on any power toying with the idea of initiating expansion by military means. What real value the intended gains will have when and if they are actually achieved is similarly far from certain. The prospects of capturing Western Europe intact cannot be rated as high by the Soviet Union, and a Western Europe that had been fought over—particularly if nuclear arms had been used—could well fall far short of making good the devastation that the Soviet homeland itself would probably have undergone. The possibility of escalation is inherently uncertain, since it is not subject to control by one side alone.

A calculus of deterrence today—by almost any concept of rationality that can be imagined—must inevitably lead to a quick rejection of the vast bulk of conceivable conflicts between the super powers. When military options are analyzed for safeguarding old gains or seeking new ones at acceptable levels of costs, risk, and uncertainty, the greater appeal of nonmilitary patterns of action can hardly be denied. The idea of Soviet military adventures in distant lands that could endanger the Soviet homeland itself was put to a salutary test in the Cuban missile crisis. Where adequate and credible military power would be encountered, i.e., where a well-designed deterrent exists, Soviet initiatives seem likely to be kept within a low level of provocation. Soviet leaders may tend also to take a greater interest in the use of proxies—e.g., the Cuban forces used in Angola.

Viewed from the standpoint of the West, deterrence poses continuing questions as to its proper level and content. It is hard to determine just how much is required.

A full deterrent might approach a full capability for defense. It would deny an enemy the prospect of territorial gain, exploitation of resources, or extension of political control in ways adverse to the West. A further Western capability, actually to prevent large-scale military damage to our homelands, would have the value, if it could be achieved, of denying to the Soviets the ability to use this potential for damage and destruction as a basis for exerting threats for purposes of coercion or intimidation.

Such capabilities lie beyond the realistic, however. In Western Europe, for example, it is not within the power of military forces to prevent all seizure of land, nor is it possible to prevent deep inroads and damaging attacks by aircraft, missiles, and submarines against NATO territory and the surrounding seas. This does not of course mean that there are no possibilities whatsoever along these lines. NATO has a good chance of halting Soviet advances over much of its area, despite some unavoidable loss of terrain—if the NATO powers are prepared to back up conventional forces at critical stages with tactical nuclear weapons selectively employed. And the damaging effects of air and sea attacks, although not within NATO's power to prevent entirely, can be limited to some degree and within some areas. With regard to strategic nuclear threat, however,—the Soviet land- and sea-based missiles, the high-performance bombers—there is all too little that can apparently be done by direct military action to prevent or reduce the destruction they could achieve if full-scale attack should occur.

It is in the other factors of deterrence, however—the factors of cost, risk, and uncertainty—that we find much of its real strength and promise. The retaliatory power of our strategic missiles and bombers offers an impressive potential response against similar attack on the United States, and this power has important deterrent effects as well against a possible Soviet military attack against Western Europe, which might escalate to all-out conflict. The costs and risks to which the Soviets would be exposed would be very great. The introduction of flexibility through options that would enable the U.S. strategic nuclear force, if such were the decision, to be employed selectively and in a succession of controlled nuclear operations against carefully circumscribed military targets tends to strengthen its deterrent effects. The power to inflict heavy losses on attacking troops of the Warsaw Pact in Europe and on their air and logistics support activities in Eastern Europe is of central importance to the deterrent there. Obviously such operations would carry with them the risk of further and deeper involvement if the other side—the Warsaw Pact—were in turn to decide to reinforce or broaden the scale of its attack. But this added risk cuts both ways, since it extends the possibility of attack to the homelands of the aggressors, not initially the scene of ground fighting. This asymmetry between aggressor and victim—that the aggressor's own territory was initially less caught up in actual destructive combat—provides an important lever for Western efforts to restore the deterrent, to discourage further escalation by the aggressor, and to terminate the war.

The uncertainties of war are manifold, even after long combat has demonstrated the current capabilities and limitations of the weapons being utilized. The chances of battle, the possibility of unforeseen maneuvers, the introduction of new tactics, the vagaries of leadership, weather, and morale all combine to rule out any possibility of completely determining the outcome in advance. New weapons and new ways of employing them add powerfully to uncertainty regarding the course of battle from the outset. The Middle East war showed from the Arab side how new-generation antitank and antiair weapons could tilt the balance of advantage between attacking and defending forces, and the introduction of altered Israeli tactics during the actual course of the battle showed how the balance could be at least partially restored. In Central Europe, the application of these and other new weapons to the particular conditions of weather, terrain, distance, and force concentration presents deep unknowns, at least partially unresolvable, even though some benefit—probably substantial—to the defender appears to be a near certainty. The debate itself will add to the burden of uncertainty falling upon the aggressor. Similarly, the awesome unknowns concerning the effects of massive nuclear exchange must give restraint to any government that would contemplate unleashing such attacks.

Sooner or later, in setting the size and composition of military power to accomplish deterrence, we run up against the crucial but unanswerable question: how much is enough? Ultimately, the answer could only be given by the potential aggressor, but even the aggressor could not foresee the exact circumstances that would exist in all cases that might arise, or what his decisions and actions, in the actual event, would be. Any calculation will necessarily be far from exact.

In sizing our strategic nuclear power to meet the needs of deterrence, one useful broad principle is that already suggested in the examination of the policy of defense: to assure that after a full-scale nuclear exchange, the United States would survive in a position not inferior to that of the Soviet Union. Lesser forces might suffice to deter Soviet resort to nuclear conflict; in a rational world they certainly should. But inferior Western forces would tend to open the way for adventuresome Soviet tactics of threat and pressure based on its nuclear power and to a freer hand for the Soviets to operate militarily below the level of strategic nuclear confrontation.

In the same vein, the principle of parity can add to the stability of the deterrent, by denying to the Soviets in peacetime any sense of preponderance that could create an atmosphere of Soviet advantage and ultimate power to prevail.

At the level of strategic nuclear confrontation, deterrence offers a practical and effective means of satisfying major American security interests. It cannot offer complete certainty of success, since the Soviets retain their ultimate and independent power to decide whether or not they will launch their strategic nuclear force. But it can reduce to a low level the likelihood that they will do so.

Within Western Europe, and in the general area of Japan and South Korea, strong and ready forces-in-being provide a powerful deterrent to military attack and to tactics of threat and coercion. Again, just what the required minimum may be is impossible to say, prior to seeing the point at which deterrence actually failed. What is clear, though, is that as the West pares down its forces, sacrifices the ability to deny an attacker the gains being sought, and relies more and more on the costs, risks, and uncertainties that we present—which has been the long-time NATO tendency since the sense of threat began to ease—deterrence is made progressively more tenuous. No longer are there two considerations—denial of gains and imposition of costs—to drive an attacker in the same direction. The balance may be weighed between the value of prospective gains and the costs likely to be incurred. And on the Western side, the knowledge that Soviet chances of being able to seize the area are substantial and increasing could not fail to weaken the readiness to stand up to demands. This is the familiar road to "Finlandization," paved with self-reassurances that a little more unilateral cutback really will not hurt very much.

Deterrence has demonstrated, through the unprecedented thirty years of peace in Western Europe, that it can provide stability and security at moderate financial and social costs for the Western powers. Since the end of the fighting in Korea in 1953, it has worked as well in the area of Japan and Korea. The prospects are good that it can continue to do so.

Elsewhere in the world the prospects are markedly lower. The prospects for deterrence of an activist power such as the Soviet Union's seem to diminish rapidly with distance from areas of vital concern—specifically, from the respective homelands, from the centers of power and influence such as Europe and Japan, and from the storehouses of vitally needed resources such as the Middle East. The military capabilities that the Western nations pose to possible Soviet (and Chinese) intervention and penetration in such distant areas are limited in scale and becoming more so, reflecting the lower level of their security interest. The effective will of the West to oppose encroachment cannot be rated as high, particularly after its limits were demonstrated in Vietnam and in Angola. As a result, the Soviet calculation of gains versus costs, risks, and uncertainties will be affected more in such areas by their perception of the strength of indigenous nationalism and the relative strengths of internally contending parties than by what the nations of the West are likely to do. Where intervention by the Soviets would touch the security of the United States—as in Latin America, for example—the higher stake that is involved for the United States will undoubtedly add a restraining deterrent against too active a policy on their part, as it has done in the past (e.g., during the Cuban missile crisis, in 1962, and in Chile in the early 1970s).

With regard to the freedom of the seas and free passage along major shipping lanes and through narrow waters for the peaceful shipping of the West, the fundamental importance of these principles to the West and the inherent flexibility and capability of naval power to safeguard them suggest that deterrence of organized interference can be made effective. The possibilities of terroristic or clandestine interference cannot, however, be disregarded for commerce at sea nor, especially, for commerce by air, where the possibilities of protection are much more limited and uncertain.

The policy of deterrence through military means is thus more restricted in its application and in the results it is likely to achieve in relation to Western security objectives in the third world than it is in relation to the strategic nuclear confrontation, the security of major regions such as Western Europe, Japan-Korea, and the peaceful use of the world's oceans.

The constituents of deterrence—will and capability to deny gains, to impose costs and risks, and to increase uncertainty—provide the possibility of establishing a level of military deterrence that may fairly be termed "substantial." In the range from no deterrence at all to deterrence that is total or near total, the band of feasible options is today placed relatively high on the scale. The options are expensive to establish and sustain—a point that will be discussed with detente—but where they exist they limit significantly the circumstances in which military force or the threat of force is likely to be used by the great powers, particularly against each other. The deterrent options serve to reduce drastically the chances of an all-out nuclear exchange occurring, which could devastate or damage the whole world.

While we do not and cannot know whether the full forces maintained by the West for deterrence are necessary, we know that if the Soviets assessed them as insufficient, and began to move threateningly against the United States or its allies, the timely reconstitution of an adequate deterrent to restore stability would be a very uncertain thing. The period during which a balance was being restored and the Soviet advantage was being taken from them would have to be rated as one of the most dangerous the world could experience. It would make the hazards that attended their withdrawal of their missiles from Cuba seem minuscule in proportion.

Detente[a]

In contrast to deterrence, detente—the third main line that our security policy may pursue—is designed not so much to deter the threat against us as to constrain and reduce it. Our chief concern is of course detente between the Soviet Union and the United States—its implications for our security and for our military forces. Detente operates by reducing the forces poised against us (concurrently reducing our own), by reducing the causes for U.S.-Soviet conflict, by lessening the chances of uncontrolled escalation or of accidental provocations and miscalculations, and by building up a stake on both sides in the continuation of peace and peaceful commerce between the Soviet Union and ourselves.

As a key line of policy, detente suffers from a certain vagueness of meaning—and from a resulting ambiguity as to the real substance of the policy and the limits of its application. A dictionary-style definition, "relaxation of strained relations or tensions (as among nations)," does not help much toward illuminating just exactly what detente as a working concept and policy may involve.

Part of the problem, it is fair to say, may well have come from an early exaggeration and overselling of the significance of detente by some of its proponents and enthusiasts. But probably more important has been a kind of overaggregation, a bundling together into this one far-from-clear concept of so many disparate things with such diverse implications as to make the resulting policy unduly hard to comprehend and grapple with. A useful corrective is to break detente down into its four or five main segments (insofar as its military aspects are concerned), each of which then can be examined separately and evaluated with due care. The categories that suggest themselves for this purpose are: (1) the strategic nuclear force capabilities of the two sides; (2) the security arrangements that relate to such areas as NATO and Japan; (3) the security issues that involve other areas of the world, i.e., the third world; and (4) the degree of linkage that may be relied upon, i.e., the security benefit to the United States that can be expected to derive from Soviet interests in reaching and maintaining agreements on such things as economic, technological, political, and cultural activities and exchanges.

The application of the processes of detente within each of these categories offers—in varying degrees—possibilities for strengthening the prospects of stable security and world order. Each category has its own distinctive military and security implications.

In the case of the strategic nuclear forces, the SALT negotiations first addressed questions of force size and composition. Already enough has been accomplished to begin to shape and constrain future weapons capabilities. There

---

[a]Certain of the material in this section has previously appeared in my Foreword, "Interests and Strategies in an Era of Detente: An Overview," *National Security and Detente* (New York: Thomas Y. Crowell Company, 1976). I am grateful for permission to use it herein.

was a clear recognition, in the keen interest and readiness to negotiate shown by the two sides, that these are forces that, if ever used, would threaten the survival of the nation on which they were targetted. There was recognition as well that further uncontrolled, competitive expansion of the respective arsenals—whether by increasing numbers, by certain types of technological innovation, or by both—could cause instability and increased danger.

A proper early aim of the negotiations was therefore to place an overall ceiling on as many as possible of the various weapon systems—to limit the numbers of land-based, seabased, and airborne missiles and long-range bomber aircraft. While this thrust has been quantitative—that is, the limitation of numbers—the subsequent negotiations soon brought qualitative improvements, such as MIRV, increasingly under deliberation. It may even be anticipated that as detente negotiations continue through future cycles, they may be applied not only to the structure of the strategic nuclear forces (i.e., to their size and composition) but also to their levels of readiness, their dispositions, and their deployments—in fact to all elements of their force posture. If such negotiations in fact develop, the ultimate effect could well be to narrow significantly the range of strategic options available to each side for the employment of weapons of this kind—particularly those options that might tend to increase the instability of the situation, that is, alarm the other and trigger an alert, with critically short allowable response times. Likewise, under the policy of detente, continuing efforts to make firm provisions for U.S.-Soviet consultations aimed at heading off dangerous confrontations and giving opportunity to defuse disputes that could develop toward nuclear confrontation would certainly seem to be appropriate—and should have a positive security value. Measures of that kind, carefully designed and negotiated, could be stabilizing in their effects and could reduce the likelihood of resort to strategic nuclear attack and response.

In the setting of levels for strategic nuclear weapons, there is fortunately a tendency toward diminishing returns that works on the value of the higher numbers of weapons in each category. The result is to leave latitude for rough judgments of equivalence rather than attempting a detailed, rigorous offsetting of each separate characteristic—an attempt that would be foredoomed to failure through sheer complexity. The margin of uncertainty in foreseeing the exact levels of performance of specific weapon systems, as well as the resulting outcomes of specific nuclear exchange patterns, weighs against any attempts to attain complete precision.

The problem, though, is to see that the "rough judgments" that are made do not leave some crucial Achilles' heel uncovered that could jeopardize an entire force. The greatest hazard lies in technological surprises that would introduce new qualitative threats—innovations such as possible means to interfere with launching signals and instructions, to impair internal in-flight guidance or warhead separation, or to remove the present effective shield of concealment of nuclear submarines. Provided that adequate precautions are taken to safeguard against such surprises, further scope would appear to exist for application of the detente policy to strategic nuclear forces.

When it comes to major regional defense arrangements such as those in Europe (or in the area of Japan and Korea), the application of detente is less advanced. A prime example of the detente process in these areas, in which the United States has taken an active part, has been the Mutual and Balanced Force Reductions (MBFR) negotiations, underway in Vienna since 1973, testing whether the processes of mutual force limitation developed and demonstrated in SALT can also be applied to military forces below the strategic nuclear level, and particularly to ground forces. A second example was the Conference on Security and Cooperation in Europe, concluded in 1974, which covered at least a few military topics, such as major military movements and large-scale training exercises. Those provisions, even though limited in scope, nevertheless had significant implications for detente in providing a degree of reassurance—through requiring timely advance notification by each side regarding military moves that might otherwise seem threatening to the other. The reverse side of the coin was of course that they imposed constraints on defense capabilities—in particular on quick response based on unilateral decision and on the state of readiness at which military forces are maintained.

The motivation of the Soviet Union and the Western countries to reach early agreement—or indeed, any agreement at all—has been noticeably weaker in the case of MBFR than in the case of SALT. It seems to reflect a feeling within the respective governments that the contending interests are not so acutely and dangerously poised—that the potential for a swift and sudden triggering of nuclear attacks is substantially less.

At the same time, it is clear that the issues involved are still of very great complexity and long-term importance. For example, they affect the Soviet military predominance over their allies in Eastern Europe. Thus these issues are going to be dealt with cautiously and deliberately. In addition, the forces involved—ground forces, chiefly—are much more diverse than the strategic forces therefore and are harder to count and to quantify in terms common to the two sides.

It is amply evident that both the United States and the USSR consider themselves to have interests in this area that are vital to their future security and well-being. As long as those interests are confined to their respective sides of the Iron Curtain it would seem that measures of detente that tend to strengthen the stability of the existing situation in Europe may continue to find application to the security problems of the area—and a degree of support from both sides.

In setting suitable mutual levels for "general-purpose" forces—the land-sea-and-air forces that would be involved in European theater-type operations, with or without the support of tactical nuclear weapons—it is of interest that the problem is not the same as for strategic forces. The margin of uncertainty is again substantial: as new weapons not tested in combat in Europe, for example, enter the arsenals of the two sides and because the chances of war (weather, leadership failures, doctrinal surprises, imponderables of morale, resilience, and endurance) can affect outcomes so strongly. But here the higher numbers of forces do not, in general, diminish in importance, as is the case with strategic forces. To the contrary, they tend to provide a margin of advantage, which in this type of warfare can be crucial. They tend to determine, for example, who can extend furthest in a race to the sea; who has reserves to influence the battle when the other's have all been committed; and what "economy of force" areas, sacrificing space for time, may have to be accepted. Just as in the case of strategic forces, however, rough judgments of equivalence may prove to be the best approach to stable deterrence, if both sides are willing to accept this as their aim. In contrast, agreements freezing one side into a long-term position of substantial inferiority would invite pressure tactics by the other and thus threaten to undermine deterrence and stable security.

In the third areas of the world—the Middle East, for example—the basis for concrete application of the policy of detente seems even more limited and uncertain. Here the interests of the two sides, although conflicting, do not seem to result in such close and strong encounter as to create quite the same motivation toward working out mutual agreements. There do exist issues of much interest to both the United States and the USSR, but the two countries have thus far not been challenged by each other in ways that have touched directly on questions of short-term national survival. It is not yet clear to what extent and how successfully the two nations will try to work out understandings and accords and to avoid high-risk confrontation. What is clear, from past evidence, is that each will try separately to advance its interests with the countries in the areas, with the USSR probably showing a considerably greater intensity of effort, at least until the more critical interests of the United States, such as access to oil and other raw materials, begin to be challenged.

The Soviet leaders have made it very explicit that so far as they are concerned detente does not imply ideological peaceful coexistence in these third areas of the world; in other words, they wish to retain their freedom to expand. Undoubtedly, their intent is to strengthen the Soviet presence and influence within these areas and to extend the Soviet system and the Brezhnev Doctrine, if they can, in ways that will bring additional nations into what they call the "socialist camp," and keep them there.

The salient point here is that both nations—United States and USSR alike—apparently consider themselves to have a good deal of latitude for unilateral and independent operations in these regions without bringing the nuclear threat into play. As a state of nuclear parity has developed, the range of issues over which the strategic nuclear forces can be regarded as providing a credible deterrent has correspondingly narrowed. Accordingly, threats to peace and order must be expected to continue in these third areas of the world, and perhaps to increase, along with occasional pulling and hauling by both the

United States and the USSR. The application of detente and the reliance that can be placed on it for the satisfaction of our security interests in such areas thus seems likely to be sharply limited.

Regarding the high seas of the world, the United States and the Soviet Union have both shown support for constructive negotiations. Bilateral agreements for avoidance of provocative naval tactics were reached in the early 1970s, and they have resulted in a more responsible standard of behavior. Both nations are seriously engaged in the worldwide Law of the Sea negotiations—where, interestingly, a considerable coincidence of interest between the two has been demonstrated.

The final category of detente having a potential for significant bearing on our security interests, as well as significant interaction with our military arrangements, is the benefit (or lack of same) that may derive to peace and security through the linkage effect with other agreements, detente-related, that the USSR is thought to regard as important. Here obviously is a place for caution—caution not to go too far in assuming that such benefits will be very great or very dependable. It is apparent, of course, that the Soviets need and want some of the economic and technological benefits that might be made available to them from the United States under the umbrella of the detente process. Moreover, they surely want to minimize the chances of costly and risky military confrontation with the West in cases where no overriding potential gain to themselves would seem to be assured. This must give a degree of restraint to their conduct and their actions. But we should remember that even in one's own country it is a very difficult job indeed to balance security concerns against other considerations such as financial costs and economic disruption. And the U.S. experience in Vietnam tells us that a policy judgment made at one point in time may prove to be invalid later because of the gradual erosion of popular support. One must be very reserved, therefore, about assuming that restraining influences of this kind on the Soviets will make a major or trustworthy contribution to the security interests of the United States. And this, in fact, is a further reason for keeping attention focused on Soviet military strengths and capabilities rather than relying on speculations as to their intentions. We know that their intentions—their policy toward the West—could shift very sharply and rapidly when Brezhnev passes from the scene, for example. Beyond this, we must constantly remind ourselves that even the wisest among us may be mistaken in an assessment of exactly what the Soviet intentions would be when faced with specific challenges and specific opportunities at a specific future time—challenges and opportunities that neither they nor we can predict in exact terms today. Their military capabilities, which take them years to build, are much slower to change; such capabilities provide a much more reliable basis on which to calculate the forces and the posture that we should maintain.

Detente also embodies agreements establishing relationships and procedures for the easing of militant confrontation and the settlement or amelioration of disputes. Here, too, accomplishments are to be welcomed in the name of security, since their tendency is to lessen the likelihood of conflict. The West could not, however, afford to purchase such agreements at the price of one-sided force reductions that would increase Western dependence on the continuance of positive Soviet attitudes or benign intentions.

It should therefore be clear that until the leopard has indeed changed his spots, detente must be backed up by deterrence and by a strong defense. In that respect it is noteworthy that the Soviet Union has shown a good deal more skill and sophistication in combining a policy of detente and a policy of defense strength than have the Western nations. There has been no sign of complacency or of slackening in the Soviet defense program. To the contrary, their defense budgets have risen, in real terms, in every year since 1964. By the mid-1970s they were maintaining a military budget that was in real terms well in excess of that of the United States, on an economy one-half as large.

While the application of detente to defense arrangements and to security needs thus differs widely as to strength of motivation, scope, and grip of the important issues (depending on the weapons and the areas of the world concerned), the policy nevertheless has shown that it has a useful role to play when carefully tailored to the circumstances. It can serve to reduce the threat—specifically through agreed reductions in the size, composition, and readiness posture of an opponent's forces—although normally only in exchange for reductions, or ceilings, for one's own. The negotiations are inherently complex and difficult, especially since the forces, deployments, and strategic circumstances of the two sides will not be symmetrical, e.g., with respect to the location of population concentrations, critical areas, and the territory of allies.

But limited as the foregoing discussion suggests, and applied with the aims of stabilizing a broad balance of forces and reducing them by carefully balanced steps, detente provides a useful major line of policy for the fulfillment of security interests. Whatever the name by which it is known, it seems certain to remain an important part of our security policy repertoire.

5

Military Tools for the Job— Fashioning the Forces Needed

Security interests and security policies provide a basis for determining the size and composition of military forces. The process requires two further steps:

> categorizing the forces according to meaningful operational roles and assigning responsibilities to them from the security interests and policies previously examined

> determining a specific force posture for each of the major forces thus categorized

Three main groups of American military forces may be distinguished to respond to our national security interests and national security policies. They are interwoven into a complex fabric. They are (1) the strategic nuclear forces, (2) the major regional forces, and (3) the forces for dealing with problems of the third world, including free and safe use of the seas.

The strategic nuclear forces respond above all to the direct nuclear threat of the Soviet Union against the United States, but in a parallel way to the similar threat against our principal allies. In addition, there is implicit in the Non-Proliferation Treaty some U.S. responsibility for extension of a nuclear umbrella over other countries that are willing to forego the development of their own nuclear weapons. Also, in case there should occur an overwhelming Soviet attack by conventional arms against the territories of our major allies—NATO Europe and South Korea or Japan—the U.S. strategic nuclear force would be the ultimate backup in case the collective allied forces—conventional and tactical nuclear—should prove insufficient. And should the United States intervene by force in other areas of the world where critical U.S. interests were involved, the U.S. strategic nuclear force provides an ultimate backup to dissuade or, in an extreme case, to respond to Soviet use of force to thwart our purposes.

The strategic nuclear forces thus serve U.S. security interests in each of the three domains cited—relative to the Soviet Union, to our major allies, and to the world beyond. They respond, in addition, to the three policies of defense, deterrence, and detente.

The major regional forces of the United States—in the NATO and Japan/ Korea areas—are joined in common defense preparations with the military forces of our allies in those regions. The military association the United States maintains is at its closest in the case of common action with the forces of South

Korea and likewise with those of the NATO countries that participate fully in the integrated commands (i.e., the NATO nations exclusive of France since 1967 and Greece since 1974). Even with France and Greece, as with Japan, however, there exist significant measures of coordination and cooperation. The principal role of these joint and allied force commands is to provide military security within an area limited almost exclusively to the regions concerned—i.e., to the NATO region, which includes Europe, North America, and the North Atlantic, and to Japan, South Korea, and the adjoining sea-air areas.

The regional forces thus respond primarily to U.S. interests in the regional domain and at the same time to Soviet pressures on these regions. They relate primarily to the policies of defense and deterrence, although detente (through MBFR, for example) is not wholly neglected.

American forces for dealing with problems in the third world or on the high seas have a role that is far less clear—particularly in the case of problems involving third-world nations. Major ambiguities persist. On the one hand the possible adverse consequences of American withdrawal from military responsibility do not appear to have been fully recognized by American public opinion or political leadership. Nor, in all probability, have they been fully accepted, since it is very doubtful whether such withdrawal is consistent with evident American expectations of orderly foreign trade and widening international interdependence. Neither would such withdrawal give promise of denying continuing Soviet expansion into third areas of the world—expansion which in the case of Latin America, Israel, or the Middle Eastern oil-producing states would seem likely to arouse intense public concern and demands for action by the U.S. government.

On the other hand, the evidence of strong and deep opposition to any new involvement of American military forces abroad has been unmistakable since the turnabout in U.S. policy toward the Vietnam war in the late 1960s. Yet even here the picture is not wholly clear. The uneasy reaction to the powerlessness demonstrated by the United States during the Angola crisis in 1975-1976, and the favorable mood of the U.S. Congress toward the first full post-Vietnam military budget (for 1977) give evidence that this is likely to be an area of continuing flux in relation to public attitudes and public concerns.

The forces applicable to these tasks respond to U.S. interests in the third world, particularly as these interests are affected by Soviet activities. Only the defense and deterrence arms of security policy are really relevant (and even these are subject to a good deal of doubt), since detente has thus far played little role in these areas.

These then are the main categories of military forces for the security role—strategic nuclear forces; regional security forces; and forces for the "world beyond." To determine in more definitive terms what should be their size, composition, and condition of readiness, it is necessary to lay out just what military tasks the forces should be prepared to perform, what strategy they should follow, and at what strength they should be maintained. These considerations are of course closely intertwined.

They must, moreover, be meshed with factors internal to the military establishment, which also bear significantly on the determination and evaluation of force design. Before discussing the three types of military forces, it is useful to consider the criteria that tend to govern the determination of the functions and structure of our military forces. Three are likely to be of principal importance.

The first criterion is effectiveness: the results produced or producible by those forces. The effectiveness of the forces is to be measured in their positive (or negative) contributions to the prevention of war, the strengthening of a durable peace free from outside interference or coercion, and the development of more stable, less bellicose relations. Effectiveness also can be measured in terms of its contribution to the calming of potential conflict situations; to the resolving of crises and disputes on acceptable terms and without uncontrolled or unsought escalation and expansion; and to defending ourselves and our interests—or being prepared to do so—without undue losses, costs, and risks if war should nevertheless occur. *Effectiveness* is thus the measure of the security output of our armed forces; it covers, in short, all the benefits to our security expected to derive from the existence and the activities of our military power.

Economy, the second criterion, deals with input to the military effort. It requires the accomplishment of the military tasks, programs, and activities undertaken within the framework of national security interests and policies, at minimum cost in resources, particularly money, human resources, and other scarce items such as petroleum products or valuable land. It has the advantage of being more fully and more readily quantifiable than the criterion of effectiveness.

Legitimacy/acceptability, the third criterion, is one less well recognized. Its importance was powerfully shown in the evolution and final reversal of public attitudes regarding Vietnam. It relates on the one hand to conformability to recognized American political principles, social values, and institutions and on the other to the likelihood that favorable reception and supporting attitudes will be evoked from Congress, opinion-forming groups, and the public at large. There are inherent difficulties—whereas much of the body of American attitudes affecting military issues is stable, enduring, and within established perspectives, much is decidedly not—and legitimacy and acceptability are subject to volatile, out-of-context fluctuation. In any case, it is not enough for military forces and activities to give effective results, economically achieved. They must, in addition, meet these further, sometimes shifting requirements of legitimacy and acceptability as well, if they are to prove viable.

Because the military services are massive organizations with long lead times and finely tuned, closely coordinated skills, a high degree of program continuity and program duration is required. Personnel recruitment and progression must be planned not only in terms of single enlistments for short-termers, but also over a full military career for noncommissioned officers, specialists, and officers. To bring a force to high levels of expertise in the myriad technical specialties that are required in today's navies, armies, and air forces requires years for the development and refinement of techniques, for the training of specialists and supervisors, and for the preparation and promulgation of technical manuals and instructions. The accumulation of these efforts represents a priceless stock of military capital, not to be dissipated through unnecessary program fluctuation.

Similarly, research and development and procurement programs require stability and long-range planning. For a major ship, a combat aircraft, or a new tank the typical full cycle now runs twenty-five years or even more—from research and development through design and procurement, then to delivery to the fighting forces for their operational use, with continuing maintenance and modifications as required, and finally, retirement or replacement.

Changes in major force deployments and their operating concepts likewise require long years and large expenditures to be effectively implemented. Communications and command facilities must be designed and installed, depots and maintenance facilities established, and military hospitals and housing provided. The environmental, economic, and social impacts of major changes weigh further on the side of stability.

More strictly military functions such as planning and intelligence add to the premium on stability. Intelligence requirements for altered tasks must be redefined, new modes of collection (if needed) identified and created, new data banks of information accumulated, and new products provided to the users. Planners must reevaluate assigned missions, reassess the capabilities and limitations of relevant enemy and friendly forces, and revise the courses of action prescribed for our operating forces. Planning for strategic nuclear operations that requires detailed computer lay-downs of land- and sea-based ballistic missile weapons together with bomber-delivered weapons may take two years or more to respond to such changes as the introduction of a new system such as the Poseidon or Trident, or the expansion of targetting options to provide an enlarged concept of flexible response.

Yet the need for continuity and duration is just half the story. Imperatives of technological and tactical innovation and modification must also be served. The military balance the West must maintain is inescapably a dynamic equilibrium. Qualitative superiority of Western weapons has traditionally been an offset to Soviet mass, but for ten years or more the scale of Soviet research and development has been exceeding that of the West. A qualitative momentum has been built up for the future that places a rising challenge on past Western superiority.

Considerations of human resources add their further weight. Economy requires a constant search for new ways of using machines to replace expensive labor, and the imperative of reducing human losses in battle presses us to rely more on weapons and less on numbers of fighters (even though the result is to lower the so-called teeth-to-tail ratio).

Thus the dilemmas are real between the need for stability and long lead times on the one hand and the vigorous, flexible reach for new technological and tactical possibilities (as well as response to changed security interests, policies, and tasks) on the other. It is a problem that must be dealt with by flexible planning and programming systems of high professional competence and by the ability of higher civilian authorities—in the Executive and Congress alike—as well as the informed public to grapple with the issue in an ongoing way. For even while continuity and stability are being sought, constant change—contraction, expansion, revision, redirection is characteristic as well in every aspect of force design and development.

The one thing we can be sure of in proposing force strengths and programs is that they must be reviewed, revised, and re-budgeted on a regular and frequent basis. The proposals set forth in the following sections are certainly subject to this process.

6 Strategic Nuclear Forces

The Major Opposing Forces

The principal U.S. strategic nuclear forces consist of the so-called Triad:

land-based ICBMs—the ~~100~~ *1000* silo-protected solid-fuelled Minutemen and the 54 older, larger, "soft" Titans

the sea-based SLBMs—656 solid-fuelled Polaris and Poseidon missiles carried in 41 ocean-going nuclear-powered submarines

long-range, land-based bombers—463 B-52s and 66 FB-111s

Certain other nuclear-capable forces could be used in a "strategic" role[a] (and might be in particular circumstances), such as carrier-based aircraft, shorter-range ballistic missiles, and even aircraft that are primarily "tactical," such as the F-4s and the F-111s based in Western Europe. In similar fashion, the strategic forces are capable of being used in what may be considered a more tactical role—for the attack of military targets in Warsaw Pact countries, for example, in connection with the land-air battle in the NATO European area, under the command direction of NATO's Supreme Allied Commander, Europe (SACEUR).

But the forces that are of primary concern to this analysis are those first listed, and the use of primary concern is strategic: attack of an enemy's war-making capacity—industry, transportation, and human resources and, as the term now tends to be used, the military establishment itself, including, in particular, strategic nuclear offensive and defensive forces.

The essential element of the strategic nuclear forces is the thermonuclear warhead, of varying yields in terms of kilotons or megatons of destructive force, deliverable on target with varying accuracies, in varying numbers of minutes or hours after decision to launch them. The particular weapons vary greatly in other characteristics as well—chief among them their own destructability, their reliability, and their ability to penetrate to the target through enemy defenses. All these in general have to be reflected in the planning of operations and in compensation for expected attrition and performance-failure, but it is warhead yield, accuracy and time of detonation, and the number of arriving warheads by type that determine the destructive effect at their destination on the enemy's war-making power.

[a]This depends on the definition given to *strategic*—a refinement not essential to our analysis.

81

The "MIRVing" of U.S. strategic nuclear systems—the application of multiple, independently targetable reentry vehicles to land-based and sea-based ballistic missiles—has profoundly affected and increased the aggregate effective destructive power of our missile forces. Numbers of individual warheads per missile have been increased from three to ten times, although at the obvious cost of accepting reduced individual warhead yields, as well as some constraints on accuracy and on flexibility of selection of targets by geographical location and of time of detonation.

In addition to the present Triad of force types (ICBMs, SLBMs, and bomber aircraft) and their new generation successors now emerging (M-X land-based missiles, Trident submarine systems, and the B-1 bomber), four other weapon-systems are now at a point of technological availability: an air-launched ballistic missile, a mobile—or movable—land-based ICBM, and air-launched and sea-launched cruise missiles of medium to long range. Major decisions lie ahead regarding the incorporation of any of or all these systems into our operational forces during the coming years.

On the Soviet side, the strategic nuclear force Triad parallels that of the United States but with significant internal differences. Primary Soviet emphasis has been placed on their ballistic missiles, both land-based and sea-based, with less emphasis on bombers. A multiplicity of types of ICBMs is particularly evident: the SS-7 and SS-8 heavy missiles dating from the early 1960s deployed at both hard and soft sites; the very large SS-9, of the late 1960s; the SS-11 deployed in large numbers in the late 1960s and early 1970s; the solid-fuelled SS-13 deployed since 1969; and four newer missiles—the SS-16, SS-17, SS-18, and SS-19—the first of which may be intended for a land-mobile role, the last three assessed to be capable of carrying four, up to eight, and six MIRV warheads, respectively. The SS-18 missile, by virtue of the destructive power of its single warhead version and its heightened accuracy, is considered by the U.S. Department of Defense to be "capable of destroying any known fixed target."[b]

In total numbers, the Soviet ICBMs in 1976 stood at 1,500 and their SLBMs at 850, for a total of 2,350. Under the Vladivostok agreement, up to 1,320 of their land-based and sea-based strategic ballistic missiles are allowed to be MIRVed.

Soviet SLBMs are carried on five classes of submarines, the oldest dating back to the early 1960s. The newest—the long Delta, with 16 missiles of 4,200 nautical-mile range—was operational in 1976.

The Soviet intercontinental force of BISON and BEAR bombers has been stable at about 150 aircraft for more than 10 years. Traditionally this force has received less emphasis on the part of the Soviets than have the ICBMs and SLBMs. Long flight times mean that their weapons could only reach the United

---

[b]*Department of Defense, United States Military Posture for FY 1977*, Chairman of the Joint Chiefs of Staff (General George S. Brown), p. 34. Details of Soviet ICBMs are provided on pp. 32-35 and SLBMs pp. 36-39.

States long after Soviet missiles had arrived, assuming they were not launched beforehand. (And such earlier launching would risk giving the United States many additional hours of warning.) Now the variable-wing BACKFIRE is being added to the Soviet long-range aviation, with significant intercontinental capabilities, although there are differing Western views as to what its actual role is likely to be. It is, however, a versatile, highly capable aircraft, numbering fifty as of 1976.

Not much is publicly known about Soviet technological development of such systems as air-launched ballistic missiles, or air- and sea-launched cruise missiles of extended range. The low priority given by the Soviets to large intercontinental bombers suggests that air-launched missiles of extended range—ballistic or cruise—may not play a very important role in the makeup of their forces, at least in the near term.

Soviet interest in a land-mobile ballistic missile—initially of intermediate range but of possible intercontinental range in the future—is seen in the SS-X-20, derived from the SS-16.

The antiballistic missile (ABM) defenses of the two countries have the capability of providing a degree of protection for designated target areas of limited scope in each country. Action by the U.S. Congress resulted in the inactivation in early 1976 of the United States' only system—the SAFEGUARD system—deployed to protect Minuteman ICBM silos. The Moscow ABM system continued in existence as the only ABM system in an operational status.

The Air Defense systems of the two countries provide a capability of counteraction to foreign bomber attack. Reductions in U.S. forces in this category have left only very limited capabilities against any Soviet attack. The Soviet surface-to-air and interception forces, numbering many thousands of launchers and several thousands of interceptor aircraft, provide defense capabilities that are substantial in relation to the aggregate U.S. strategic bomber force. Nevertheless, the net U.S. bomber contribution as part of the Triad, even after making full allowance for the toll imposed by Soviet air defense operations, would still be of vast destructive power. Of by far greater significance than such defensive weapons in terms of the potential for destruction of strategic nuclear offensive weapons are the offensive forces of the other side. It is these that must be primarily considered in assessing the adequacy and stability of offensive force levels and deployments.

While the various weapons-types that comprise the strategic nuclear forces of the two powers are in total numerous, it is worthy of note that they are not unlimited. Each system represents ten or more years of development and of prior decision as it reaches operational status. The "menu" for choice is thus short and specific at any given time, and the combinations, for a period of ten to fifteen years ahead, are susceptible to concrete, detailed analysis as to their impact on the "assured destruction" capabilities of each side and on the stability of the deterrent in crisis conditions.

The War-Making Role of Strategic Nuclear Forces

At the most elemental level, what strategic nuclear forces can do is simply to destroy—to destroy people, industry, wealth of all kinds, and enemy military forces together with their supporting establishment. In all cases, they can destroy some but not all—a point of special significance in the case of the relatively invulnerable submarine-based ballistic missile forces at sea, for it means that each country—despite anything the other could do—could still destroy many tens of millions of the other's citizens. Although estimates vary and uncertainties abound, the strategic nuclear forces of the late 1970s have the assured capability of destroying 20 to 30 percent of the opposing nation's population (and perhaps up to 50 percent or even more), and well in excess of 50 percent of total industrial capacity. While Soviet Civil Defense programs could conceivably, in certain circumstances, substantially lower the probable loss of life, the likelihood of such scenarios occurring, on a dependable basis from the Soviet point of view, would have to be assessed as very low in any practical sense. The civil defense program does, however, add one more complicating factor, thus offering a potential for miscalculation in crisis circumstances and therefore destabilizing in its overall tendency.

Past concepts of "winning" a large-scale war have had to be drastically revised in the context of strategic nuclear exchange. Our strategic forces, along with the Soviet forces employed in such an exchange, can be the primary determinant of:

the damage to the USSR (which Soviet forces limit, but cannot eliminate) and therefore the Soviet population and industry that remain

the damage to the United States (which our forces limit but cannot eliminate) and therefore the U.S. population and industry that remain

the relative strengths of remaining Soviet and American population and industry

Neither the U.S. nor the Soviet strategic forces provide in the traditional sense an ability to hold ground—of their own country, of the enemy, or of their respective allies—nor to control or deny the use of the high seas, although the results of a strategic exchange could certainly affect the ability of other military forces (ground, sea, and air, possibly using tactical nuclear weapons) to do so.

In the circumstances of a strategic nuclear holocaust of the dimensions described, it is hard to assign any convincing or satisfactory meaning to the term "winning the war" as heretofore conceived. We are left with such criteria as bringing the strategic nuclear exchange to an end, preventing the imposition of alien rule, and providing the opportunity for national rebuilding, as free as possible from the threat of resumed attack. The difference between "winning" and "not losing" could be solely the imposition of constraints on the enemy that gave us assurance that hostilities would not be resumed or national freedom from outside coercion or the economic welfare of our people be jeopardized further.

But we must not mislead ourselves into the kinds of calculations that equate enemy losses to our own insofar as security aims are concerned. Our true aim is to protect our people, not to destroy others. Enemy losses compensate for our own only in the narrow sense that they make possible an outcome in which we would not be at a continuing disadvantage, or subject to foreign domination. In effect, we would have paid the price of trying and failing to prevent war while pursuing other values in a situation of crisis and confrontation, and then conducting the war in a way to achieve the most advantageous, or least disadvantageous, possible outcome.

The condition and evaluations just reviewed are repellent and mind-numbing even to attempt to consider. But they must be faced, and should be understood, since they are not only the ultimate reality in case central nuclear war should ever occur, they are also the strongest support for the courses of deterrence and detente—the dissuasion of war and the reduction and partial control of the likelihood of its occurrence—and also for the avoidance of actions in other spheres of encounter that generate the risk of escalation into nuclear war.

Security Interests and Policies Served by Strategic Nuclear Forces

The interests, policies, and task objectives that our strategic nuclear forces should help to serve have been set forth in earlier chapters of this book, together with the criteria that should guide the design of such forces. It is useful to recapitulate them briefly here.

The relevant national security interests include:

1. denial to the Soviet Union of any possibility of using its strategic nuclear forces to advantage through initiating an attack on us. We hold this powerful destructive force in check while assuring that our other great values are safeguarded and kept free of impairment from outside forces.

2. denial to the USSR of a position of military strength and advantage from which they could by threat of military attack dictate terms to the United States—that is, determine the course of international affairs.

3. denial of territorial aggrandizement, of seizure and control of free world resources, or of extension of their influence and control into foreign lands that would carry serious hazard to the United States. Western Europe and Japan and the oil-rich areas of the Middle East are high on the list of such crucially important areas.

The major security policies of defense, deterrence, and detente impose further requirements on our strategic nuclear forces. Requirements for defense include the following.

1. Giving us a reasonable expectation, in case of all-out nuclear exchange, of being able to terminate it in a position not inferior to that of the Soviet Union.

2. Providing a capability of conducting counterforce attack against Soviet nuclear forces and other military targets (as they can against us) with results not to our relative or cumulative disadvantage.

Requirements for deterrence include the following.

1. Providing assurance that after full-scale or limited nuclear exchange we could survive in a position not inferior to that of the Russians.

2. In the process, posing to the Soviets such prospects of denial of gains together with costs, risks, and uncertainties, as to make the perceived relation of losses to benefits of resort to military force, or the threat of force, unattractive to them.

3. Reinforcing the deterrent by denying to the Soviets—as to ourselves, our allies, and the neutrals—any sense of Soviet preponderance that could create an atmosphere of Soviet advantage and ultimate power to prevail.

4. Reinforcing the deterrent further by designing forces that minimize the premium for—or compulsion toward—surprise or preemptive attack against vulnerable but threatening enemy forces—i.e., providing forces that strengthen crisis or operational stability.

Requirements for detente include the following.

1. Establishing agreed limits on the numbers of strategic nuclear weapons of all important types, at the lowest possible levels that will satisfy the needs of defense and deterrence just stated.

2. Establishing agreed measures that will limit or control the introduction of new technological advances in ways that:

continue to satisfy the needs of defense and deterrence

avoid competitive circumvention of quantitative limits by qualitative improvements

suppress technological innovations that would have an overall destabilizing effect

capitalize on the "diminishing returns" effect of higher numbers of strategic nuclear weapons, by which target-damage by additional weapons drops below that to be expected from the initial attack

make sure that no "Achilles-heel" is left uncovered that could jeopardize an entire force or any one of the Triad's components (i.e., the ICBMs, the SLBMs, and the bombers)

The broad task-objectives to be served by our strategic nuclear forces include:

responding to the direct nuclear threat of the Soviet Union's strategic nuclear forces against the United States, and against our principal allies

providing a nuclear umbrella protecting other countries that forgo development of their own nuclear weapons

providing the ultimate backup in case of overwhelming Soviet attack by conventional arms against the areas of our major allies (NATO Europe and South Korea, for example) in the event that the collective allied forces in the area (conventional and tactical nuclear) should not suffice

providing a backup to dissuade or, in an extreme case, to respond to Soviet use of force to thwart our purposes, should the United States intervene by force in other areas of the world where critical U.S. interests were involved.

The foregoing list of requirements defines a heavy but not an impossible burden for our strategic nuclear forces. The size of the burden is illustrated by the size of the strategic forces' annual budget. Typically these forces in recent years have cost no more than 10 to 15 percent of the total annual defense budget to develop and maintain—$9.4 billion in direct costs in Fiscal Year 1977, no more than half that much again as the "allocated" share of establishmentwide costs, for a total in the range of $15 billion. The success achieved and achievable in terms of fulfillment of our prime security objective—deterrence—has been well demonstrated. In more than thirty years since Hiroshima and Nagasaki no third nuclear weapon has been fired with hostile intent. And there is rising general confidence on the part of the general public—paralleled by strengthened expectation to the same effect among governments—that this condition of nonuse, at least in terms of strategic nuclear exchange by the military superpowers, can continue.

The strategic nuclear requirements broadly defined above center on the overriding principle, or basic option, of denying to the Soviets a position after nuclear conflict that is superior to our own.

At various times during the nuclear era, and from various analysts of the nuclear equation, several alternative major strategic nuclear options significantly different from the above have been identified and proposed. These alternatives would be carried out in war and prepared for in peace (with the hope that war will be prevented).

The first such alternative option would set our national sights higher, and aim at nuclear victory—an actual war-winning objective. But because the USSR has now developed a massive nuclear arsenal of its own, and because the damage to our people and our country in case of actual large-scale nuclear war would inevitably be so severe, even the mention of *victory* has an incongruous ring. The difference between *superior* and *not inferior*, can be significant. A perceived aim of emerging from such an exchange in superior position to our adversary carries risks of setting off a competitive and destabilizing reach for stronger aggregate nuclear forces. Unless some natural factor were found that could be made to penalize the USSR more than ourselves because of their aggression, superiority seems an unrealistic aim for the United States to set for itself. In any case, as suggested earlier, "victory" might mean little more than a measure of dominance by a badly battered United States over a worse-battered Soviet Union.

A nuclear option at a lesser level than the one we have chosen (of denying nuclear victory to an attacker) would be one limited strictly to heavy assured destruction of enemy population and industry. In the 1960s, such planned destruction was set by the U.S. Department of Defense at 25 to 30 percent of Soviet population and 50 percent or so of Soviet industry. No one can know, of course, whether such devastation would destroy a nation's ability to survive as a self-sustaining society. Its chances of doing so without the subsequent loss of even greater parts of its population could not be good, however. And in any case such a prospect must exert a strong deterrent. The option presupposes a nuclear posture that could ride out a Soviet nuclear attack, whatever its form, with enough of our own nuclear forces surviving to accomplish the designated levels of destruction. It is vulnerable, however, to the possibility (discussed later) that an enemy having a hard-target capability (as the Soviets have been developing against our 1,000 Minuteman force) might conduct a counterforce operation that—if we limited ourselves to urban industrial attack capabilities and no counterforce weapons—could leave us with the choice of doing nothing or launching the mutual urban/industrial holocaust. Also, this option omits the requirement to end up not inferior to the adversary—thus leaving open the possibility that the enemy might be sufficiently superior to impose control upon us.

An even lesser variant of the assured heavy destruction option would be very limited destruction or demonstrative use. Some, with the Hiroshima and Nagasaki precedents behind them, argue that merely the capability on our part of destroying several Soviet cities would be adequate to dissuade the Soviets from attacking us with nuclear weapons. Elements of this concept are to be found in French views regarding the role of their nuclear forces, and in British statements reserving the possibility of unilateral use of theirs for supreme national interests. If the Soviets, having much larger nuclear forces, were nevertheless to launch a nuclear attack, then—even if we were actually to employ our limited force—they could continue on to destroy and dominate us. And such a limited force would offer little toward other task objectives set forth earlier, such as providing a nuclear umbrella for the nonnuclear nations, backing up NATO, or standing in support of U.S. operations elsewhere.

Demonstrative use—the explosion of one or a very few weapons, perhaps against significant targets, perhaps in an isolated, "no-target" area—suffers from some of the same deficiencies. It has received considerable interest in NATO, over many of its years, although always in the context that if limited demonstrative use did not suffice, much larger weapons inventories would still be available to be used in more telling ways. Even so, the concept has always been open to the charge that it risks demonstrating to the enemy not our resolve but the limits of our resolve, with counterproductive result on our efforts to end the war. Although the point is controversial, the self-imposed limits observed by the United States in the early years of bombing in North Vietnam may have encouraged the North Vietnamese to believe that they could afford to continue the war. They only agreed to halt it when bombing for full military effect was finally conducted in late 1972 and early 1973. And in the present case if our forces and weapons were limited to a few for demonstrative use only, the Soviet preponderance would be virtually impossible to resist, even in peacetime, and would be so assessed by the Soviets, the United States, and our allies.

An option more recently proposed has been that of selective targeting, sometimes called "flexible response." It has highlighted the possibility of a series of exchanges, possibly of some duration, involving attacks against military as well as urban/industrial targets. The concept is not new, since nuclear planning has long incorporated attacks against military targets, particularly those comprising the Soviet nuclear capabilities, as part of a "damage-limiting" objective. A degree of flexibility was provided through the ability to select a limited number of targets for attack from the whole planned inventory and to withhold attack against the rest. But while such targeting was possible, it would have been time consuming—at a moment when time would have been in extremely short supply and when possible collateral effects on targets desired to be withheld would not have been immediately evident or readily assessable. What is new now is the specific definition and development of selective target packages and the voluminous detailed planning to make flexible choice a realistic and ready option. This approach does not conflict with the "denial of enemy superiority" objective previously described; in fact it reinforces it and rounds it out. The counterforce capabilities of the newest Soviet ballistic missiles must be counterbalanced by our own if the principle of assuring that there is no resulting inferiority after nuclear attack is to be preserved for the United States. The desire of successive U.S. Presidents and Secretaries of Defense for effective options intermediate between inaction and Armageddon is on the way toward fulfillment, without, apparently, making the whole system so complicated and cumbersome as to be unworkable by the military forces or so time consuming as to be unuseable in a crisis situation. The reported reduction of retargeting time for a Minuteman missile from more than fifteen hours to approximately thirty minutes gives effect to the proposal.

In the actual employment of strategic nuclear forces, they must be prepared for use either in a retaliatory option or in a preemptive or "first-use" mode. (To bar misunderstanding it should be emphasized that the latter is neither "preventive war" nor a "first-strike," i.e., a fully disarming effort.)

Until the introduction by the Soviets of their newest ICBMs (specifically the heavy, highly accurate SS-18), our Minuteman forces could be protected for retaliatory use by the hardened silos in which they are deployed. The new threat means that reconsideration and extension of protective measures will be necessary. The Polaris and Poseidon submarines when at sea are practically invulnerable to enemy first attack, and the Trident, with its much greater missile range, will reduce even further the already low possibility of enemy detection and attack of our SLBMs by attack submarines prowling the narrower reaches of the deployment areas and routes currently in use. For our bombers, quick takeoff on warning together with dispersal and airborne alert in time of tension offer effective means of preserving the force.

The option of preemption or first-use offers the means of striking enemy missiles and other nuclear targets in their deployment sites and of avoiding being struck in ours. There can be no assurance that such employment, particularly if on a sizable scale, will prove practicable. Yet the capability adds to the deterrent by telling would-be aggressors not to be sure of launching *all* their force with first-use effect, i.e., free of any possibility of suffering significant attrition themselves, or free of any risk of failing to catch our forces on the ground. A first-use capability on our part does introduce some destabilizing tendency, but this effect is limited by the Soviet ability (similar to ours) to maintain an enormously powerful and invulnerable SLBM force at sea.

A first-use capability against selected targets, limited in number, is an inherent requirement of the flexible use option. From time to time, proposals have been made that the United States should adopt a "no-first-use" doctrine. Such would, however, weaken the deterrent as suggested earlier, and detract from the support given by the existence of our strategic nuclear forces to deterrence in the NATO area (or defense if conflict should occur) and to the safeguarding of possible U.S. operations or activities elsewhere (in support of Israel or for the protection of access to oil, for example) against Soviet interference.

Principles Governing Force Size and Composition

For determining the specific size and composition of our strategic nuclear forces, three main principles, each relating our forces to the size and capabilities of Soviet forces, appear to merit consideration. They are "rough equivalence," stability, and equitable reduction. If carefully applied, they appear to provide for the tasking options, security policies, and interests previously identified and discussed.

The Need for Rough Equivalence

Rough equivalence as a principle is open to a certain measure of controversy. It has the value of being practical, understandable, and naturally persuasive. It raises two questions, however: Could we get along with less? Can we in fact determine rough equivalence, given that the nuclear systems of the USSR and the United States differ so widely as to types, numbers, and yields of weapons, as well as accuracies and many of the other characteristics that determine their destructive capabilities?

Whether the United States needs strategic forces roughly equivalent to those of the USSR is a matter on which reasonable and responsible people who are well informed in these matters can and do differ. One alternative approach, certainly not without merit, would be to calculate the targets in the USSR—military and urban industrial—whose destruction would meet the security needs and objectives of the United States should strategic nuclear conflict ever occur. Such objectives may be stated, for example, in terms of the percentages of total urban industrial strength that should be destroyed, as well as the total proportion of military targets by type. These would constitute "assured destruction" objectives.

Given the lists of targets, techniques exist by which to apply factors of accuracy, yield, cross-targeting, attrition, operational unreliability and the like, to determine the number of weapons of various types that would give a specified probability of inflicting a specified level of damage on a specified percentage of such targets by type. Some of those in a position to know believe that this number of weapons is likely to be considerably below the number we now have, the 2,400 total systems contemplated by the Vladivostok SALT accords, and the 1,320 subtotal of allowable MIRVed systems.

The difficulty with this approach—a difficulty that the proponents themselves recognize in varying degrees—is that such lower U.S. numbers, if the Soviets maintained the full SALT allowances, could leave the United States substantially and visibly inferior to the Soviet Union—by many hundreds of launchers, and several thousands of warheads. A natural rejoinder would be that if the Soviet Union wants to waste its substance on weapons that are redundant and worthless, that is no reason why we should do likewise. But this argument has several defects. No one can say that no value whatsoever would accrue to the Soviet Union from the extra weapons. The new targeting doctrine of selective use—against military targets for example—shows how the increment might be used by them to advantage. The United States, with lesser numbers, could be *forced* earlier to the choice between inaction and an escalation to urban industrial exchange. And the perception of this disadvantage would be harmful to Western resolution. Second, the added numbers would permit the Soviets to attain a higher destruction level against our people, our cities, and our productive capacity than we could against theirs—instead of destroying a fourth of our total population, as we might of theirs, they might destroy a third or even a half of ours, leaving them a large numerical advantage in a postwar period. Further, if technology should increase the hardness of enemy targets, or reduce the survivability of our own forces after the initial calculations were made and the levels of our forces were set, it would be necessary to increase the strength of our forces in relation to theirs, thus introducing instability and the seeds of a possible new arms race.

But overarching all these arguments is the question of the political and psychological sense of inferiority that would result. One need only recall the near hysterical overtones in the United States over Sputnik and the cries of "bomber gap" and "missile gap" in the late 1950s (both in fact nonexistent) to realize how volatile American opinion can be to charges of this kind. In the late 1970s, when deep-seated unease exists over national failure in Vietnam and impotence in Angola, at a time of lowered national confidence and consensus, but heightened suspicion and disparagement of authority, such inferiority would be an unwarranted burden on calm and resolute judgment and a dangerous tool to hand to the demagogues, detractors, and sensation-mongers.

The impact abroad of the picture of an America so disadvantaged could not fail to be profound. To suggest only one such effect, the base of confidence that underlies the stability and cohesion of Western Europe would be impaired. This confidence—so painfully rebuilt in the early and middle fifties—would certainly suffer, and the possibility would open up of a more serious exposure to the pressures and processes of Finlandization for the nations of Western Europe.

No countervailing benefit from such a course is readily apparent. The increment of cost involved—a minor part of the strategic forces budget, which is itself but a minor part of the total defense budget—is of small proportion in relation to the detrimental effects that can be foreseen.

But if the principle of rough equivalence is to be adopted, will we in fact be able to determine how to translate it into definite numbers of specific U.S. and USSR weapons? As earlier discussion has indicated, these are greatly dissimilar in key characteristics such as yield, accuracy, and range, not to mention the relative emphasis that has been placed on bombers versus missiles and on the inclusion of missiles with cruise in addition to ballistic trajectories. If it were necessary to match identical items with each other, or even class against class within the two countries, the task would obviously be impossible. At best, it is by no means easy.

Many factors apply, but there are two or three in particular that have substantial importance and at the same time provide us with a degree of elbow room in making calculations of the kinds required.

The increase in destructive effect achieved by nuclear weapons does not match increases in yield. Because of the applicable scaling laws, doubling of yield only increases the radius of a given destructive effect by something like 25 percent and the area by something like 50 percent. Moreover, for the same reason, against "hard" targets, increases in accuracy can substitute with great advantage for increases in yield. Reductions in CEP (circular error probable—the radius at which as many rounds may be expected to fall inside as outside the circle) have a multiplied effect in increasing the expectation of damage. Halving the CEP more than doubles the expected damage.

In addition, the higher numbers of weapons, where they simply result in adding more weapons of the same kind to a given target, are subject to a law of diminishing returns. If, when all factors of reliability, penetration, accuracy, and yield are considered, a first weapon gives 75 percent probability of achieving a specified degree of destruction, a second weapon of the same kind may add an increment of less than 20 percent, and a third may add less than 5 percent. Of greater importance than simply increasing the number of weapons of a single type planned to be placed on a single target is the application of cross-targeting, that is, planning to strike the target with a second, different kind of weapon—a sea-based Poseidon missile, for example, which is not subject to the same type of possible operational loss, failure, or attrition as a first land-based Minuteman.

Finally, the specifying of damage levels is far from an exact science. A shift, for example, from 30 or 35 percent downward to 20 or 25 percent as a destruction objective for urban industrial attack introduces changes in the numbers of weapons needed that swallow up many significant differences in the weapons characteristics of the two forces. And no one can say which of the two percentage levels—or others higher or lower—corresponds to the degree of damage needed to "destroy" an industrial society as a self-sustaining entity or to the degree of deterrence needed to prevent resort to war or initiation of pressure tactics based on the threat to use force.

In addition to rough numerical balance, a rough qualitative balance with Soviet strategic nuclear forces—expressed in broad operational terms—is also needed. As the Soviets achieve yields and accuracies that, in combination, give them a hard-target capability against our land-based Minuteman ICBMs we must be sure we develop and introduce the technology to do likewise to theirs. As advances in Soviet air defense capabilities threaten to "degrade" our bomber force capabilities—and the bomber contribution to the assured destruction objective—offsetting measures such as the introduction of the Air-Launched Cruise Missile become necessary.

The Need for Stability

Such technological steps and countersteps have an inevitable tendency of disturbing the stability of the strategic force relationship—the second guiding-principle for the design of our strategic nuclear forces. The term *stability* has two main applications in this connection. The first refers to maintaining reasonably stable overall force levels—avoiding increases in force levels first by one side and then by the other in an unending effort to get ahead (or even to catch up to perceived advances by the other side).

It is one of the strengths of detente—the process of negotiation—that it offers a means of putting a lid on such reciprocal efforts—quantitatively, in the first instance and qualitatively more incompletely and uncertainly. This type of stability is commonly termed program stability; it could be called "force structure stability."

The second application of the term is to a stable condition in the sense that major force components of great destructive power are not exposed to disarming attack. The effect of such exposure—classically illustrated in large numbers of soft ICBMs or IRBMs targeted on the other party—is to raise invitingly high the premium on surprise, preemptive attack, and on quick launch to avoid such attack. Concealment of missiles at sea, hardening in silos that would take a major enemy effort to try to destroy (with uncertain results), and establishment of ABM defenses of deployed missiles with the same purpose and effect—these are the typical means of reducing this form of instability, which is sometimes called "crisis stability." It is stability of operational decisionmaking in this crucial field of deterrence or defense.

The two kinds of stability are related. To overcome instability of the second sort, force improvements are amply warranted, even if they add to overall capabilities and are subject to being matched by an opponent—preferably in ways not again adverse to crisis stability. It will remain a continuing challenge for the United States and USSR to establish qualitative restraints through negotiations that will head off or contain the destabilizing effects of technological advances.

In the late 1970s and early 1980s, MIRVs, cruise missiles, and improvements in accuracy are likely to be the three principal examples of sources of instability produced by new technology. The process of dealing with such instability becomes practicable only where the new characteristics can be translated into something that is countable or measurable. By taking note which of the new Soviet missiles have been tested with MIRVs, it becomes possible to set a ceiling on MIRVs by setting a ceiling on missiles capable of being equipped with MIRVs (i.e., the SS-17s, SS-18s, SS-19s, and perhaps the SS-16s).

The cruise missiles now being developed by the United States assist stability by assuring that our bomber force does not lose its effectiveness through tightened Soviet air defenses that add to the costs and difficulties of penetrating to the target areas. But they offer (particularly in the sea-launched version) a weapon system outside the scope of previous SALT agreements, if mounted on attack submarines or surface vessels. Unless means can be found to bring cruise missiles into the aggregate strategic weapons inventories controlled by agreement, they will provide fuel for potential competitive expansion. The problem should be more easily manageable for the bomber fleets covered by the agreements, since cruise missiles will take away part of the load-carrying capacity otherwise usable for bombs. The cruise missiles (or air-launched ballistic missiles) can also be placed on platforms that have not been covered by SALT: attack submarines, surface ships, or large load-carrying aircraft in particular. These raise difficult problems of defining and limiting the scope of such applications. Verification becomes a major concern—for attack submarines, for example— since much of the fleet could be fitted out in this manner. Though difficult, the negotiations should not be impossible, however, since the missiles do not create a new, higher order of vulnerability for the opposing strategic force. In fact, they constitute merely an addition to an already massively destructive aggregate force.

Improvements in accuracy, despite their destabilizing tendencies, seem unlikely to be brought under SALT-type controls. No means of doing so has yet been devised. There seems to be no substitute in this area for maintaining vigorous research and development efforts to make sure the other side does not gain a substantial unilateral technological advantage. The push of technology in this direction—whether through terminal guidance, stellar or satellite navigation, or other new methods—seems inexorable. A by-product may be, as land-based systems in fixed silos become more vulnerable, a shift of interest and emphasis toward land-mobile systems, air-mobile systems, or the well-proved submarine systems, all offering a more stable force posture.

The Need for Equitable Reductions

The third major principle applicable to force design is that of equitable reduction. This is the task for future SALT deliberations that ought to be a continuing, institutionalized process. It should, by adding to common under- standing of the risk implications of various forces and levels, strengthen the stability of the deterrent. The Vladivostok accords, based on a 2,400 ceiling for aggregate strategy delivery vehicles and a 1,320 ceiling for missiles capable of carrying MIRVs, were sharply challenged when announced for setting the levels unnecessarily high. Quite evidently, however, these were the lowest levels on which agreement could be reached at the time. But there is nevertheless no inherent reason why considerably lower levels cannot be worked out in the future—with benefits in economy and, in all likelihood, in crisis-type stability.

It is extremely unlikely that any entire element of the Triad can or should be eliminated. The Polaris and Poseidon SLBMs have high invulnerability and add stability to the deterrent. The Minutemen in silos provide high accuracy, excellent communications, quick response and target-selection, and high assessed reliability. The bombers add heavy weight of effort, diversity of attack, and the increased assurance of systems tested in combat. They are a multipurpose force, useful to show resolution in time of crisis. Moreover, as noted earlier, the number of systems seems likely to increase rather than decrease as cruise missiles, air-launched ballistic missiles, and land-mobile (or multiple launch-point) systems become operationally available. Nevertheless, there is significant scope for mutual reductions, without detriment—in fact, with benefit—to the security interests and policies our strategic nuclear forces should serve. Aggregate reductions of 30 or even 50 percent do not seem inconceivable, although the bargaining will be long and hard and complicated by asymmetries of population location, industrial concentration, directions of approach, existing technologies and weapons investments, assurance of verification, and other factors no less difficult to appraise.

One may expect that a special limiting factor on efforts to reduce force levels will come from the Soviet side. Without doubt, they will wish to reinforce their superpower status by maintaining an overwhelming preponderance of nuclear strength over nearby neighbors—the Chinese in particular, but Western European nations and their own allies as well. Deeply aware that their hegemony where it exists rests crucially on military power, they will not wish to tempt dissidence with too narrow a superiority. And their massive nuclear power provides them an umbrella under which they can press the effort to expand the sway of the socialist system even to the endangerment of existing governments and regimes. The sense of power that their nuclear arsenal gives them is well reflected in the confidence of their assertions about the worldwide "correlation of forces," which they assess to be favorable to them, and becoming more so.

Another factor also will weigh against extensive reductions. It was suggested earlier that weapons at the higher numbers diminish in their contribution to assured destruction, for example, and thus permit rough equivalence rather than attempted precise matching characteristic by characteristic—i.e., yield for yield, accuracy for accuracy, relative speed of delivery, invulnerability, and the like. SALT agreements become more achievable at higher levels. As numbers decrease, the opposite effect will come into play, and asymmetries will be more intensively argued. A consequence may be some convergence in the respective force designs of the two countries if such reductions are achieved, notably in an increasing relative emphasis on submarine-launched ballistic missiles, with reductions applied mostly to the ICBM and bomber components of the Triad.

Implications of Soviet Force Development

But this lies in the future. Thus far, the evolution of Soviet weapons and forces has tended to require strengthened and improved, rather than reduced U.S. forces, if undiminished U.S. security is to be maintained via the routes of rough equivalence and maximum stability.

Soviet developments affect all three components of our force Triad: our ICBMs, our bombers, and our SLBMs. Soviet MIRVed ICBMs of improved accuracy and substantial yield pose a threat to the U.S. Minuteman deployment.

Methods of countering it include the development of an increased hard-target capability on our part through improved accuracy for our missiles, the development of land-mobile systems, and improvements to the survivability of the Minuteman protective silos.

Continued strengthening and thickening of Soviet air defenses have posed a potential threat to the effectiveness of the American bomber force in its strategic nuclear role. Also, the increasing Soviet numbers of the longer range (4,200 nautical mile) SS-N-8 submarine-launched ballistic missile increase the danger that many American bombers could be caught on the ground in case of sudden attack. Methods to counter these adverse effects include the introduction of extended-range air-launched cruise missiles, the development of an air-launched ballistic missile system, the development of improved penetrating power for the B-52 or the planned B-1, and the improvement of rapid-launch capabilities.

The antisubmarine warfare (ASW) threat to our submarine missile forces is low, and it seems likely to remain so. Nevertheless, Soviet capabilities are being improved, from an initially limited starting base. These efforts can be countered by further quietening of our future submarines—the Trident, in particular—and by the introduction of the longer-range Trident missile, which is expected to increase by a factor of four the operating area within which our SLBM forces can operate.

Essentials for the U.S. Force Program

Out of this analysis, it is possible to identify the main lines of development for U.S. strategic nuclear forces in the period ahead (through the 1970s and most of or all the 1980s) which would correspond to the principles, options, policies, and security interests thus far developed.

1. Aggregate forces roughly equivalent to those at the Soviet Union should be maintained, preferably governed by the SALT agreement and its successors, including the Vladivostok accords. The ceilings currently contemplated (2,400 strategic delivery vehicles with no more than 1,320 missiles of MIRV-equipped types) provide a satisfactory beginning.

2. If agreements can be achieved, substantially lower levels could be established, particularly in land-based missiles and intercontinental bombers—perhaps at two-thirds or even one-half of present numbers. However, special care should be taken to see that Soviet advantages (in throw-weight and yield, in particular) are safely counterbalanced.

3. The basic Triad of forces should be maintained, and they should be supplemented with air-launched cruise missiles and land-mobile missiles as necessary to maintain the stability of the deterrent, and rough equivalence overall with Soviet forces. Such additions should preferably be made subject to

SALT agreements. If reason develops to introduce air-launched ballistic missiles, they too should be made subject to SALT agreement. Unless means of overcoming difficulties of definition, limitation, and verification can be overcome, sea-launched cruise missiles would best be dealt with by prohibition; failing that, the United States would need to base its actions on intelligence assessments of Soviet action. (It may be noted that because of the heavy U.S. urban concentrations close to our sea-coasts, Soviet sea-launched missiles could be of much shorter range—well below the 600 km. that has been under discussion—and still be comparable in their effectiveness to much longer-range U.S. missiles.)

4. Research and development should be such as to keep us qualitatively on a par with, or preferably ahead of, the Soviet Union so long as the latter retains substantial advantages in total missile throw-weight (now running between 2 and 3 to our 1). High-priority effort should be given to progressive improvements in accuracy, and in the throw-weight (and yield) of our missiles. The Soviets must not be accorded a significant unilateral advantage in hard-target kill capacity.

5. Modernization of key weapons systems—the B-52, the Minuteman, and the Polaris/Poseidon submarines—to overcome the effects of aging and improve operating efficiency (i.e., internal needs) should at the same time incorporate the external requirements of mission effectiveness and system stability previously discussed.

A force initially configured and progressively evolving along the foregoing lines would respond well to the criteria of effectiveness, economy, and public acceptability, which must guide the determination of strategic forces role and structure for the long pull. It is a force that would exert a steady and strong deterrent on the Soviets over a wide range of possible conflict situations. It would serve effectively if hostilities should nevertheless occur, and it would encourage the Soviet Union to continue to pursue the processes of detente in the strategic nuclear field. Although expensive, it would amount to but a minor fraction of the total defense budget (i.e., in the 10-15 billion dollar range of a total that is exceeding $100 billion), and it would be economical in the sense that its main elements correspond to the identifiable needs and major constituents of a policy-directed force. While it epitomizes the risks of nuclear war, it reflects conscious effort to bring and keep those risks to the available minimum.

The specific details of the program will be difficult to determine and to negotiate because of their inherent complexity. Nevertheless, even the present analysis suggests that the essential facts can be organized and presented in understandable and reasonably simple overall terms. If the final criterion mentioned above—public acceptability and support—is indeed to be achieved and sustained, then open and honest explanation and discussion of the issues involved appears mandatory on the part of the Executive, bringing in the Congress, the public, and the interested opinion-leaders. Although the initial process may thereby be made slower and more arduous, by such widened participation the final outcome will be much more likely to receive support and the process to be permitted to continue effectively.

7 Regional Security Forces

Within the noncommunist world, the two regions that hold special imperatives for U.S. military commitments are Western Europe (with its Atlantic approaches) and the Japan/Korea complex in Northeast Asia/Northwest Pacific. The conviction of Americans that their security fate can be determined in Western Europe, well shown in two world wars, remains strong today. And the contrast is sharp between the peaceful, stable northern Pacific of the present and the pre-Pearl Harbor belligerency in the same area.

In these areas, the security arrangements and military commitments for the United States can only be examined and appraised in a collective context. In the Western European area, U.S. forces provide no more than 10 to 15 percent of the total of NATO's ground forces, about 20 percent of the ships and 25 percent of the aircraft. Even in the central region of NATO (West Germany and the Low countries) the United States provides less than 25 percent of the total NATO military manpower in peacetime. Similarly, in the Far East, the South Korean armed forces and the Japanese Self-Defense Force far exceed, in total, the American military strength deployed there. The collective military structure is far more extensive, and the degree of multilateral integration further developed in NATO than in the Far East. Yet in both areas the military problem can only be properly understood by viewing the forces as a whole. In the present study, the regional analysis is concentrated on NATO, which well illustrates the issues involved.

The principal military forces involved in regional security efforts are "general purpose forces"—the land, naval, and air forces, exclusive of strategic nuclear forces, which would conduct the battle in the area if hostilities should occur. They include both conventional and tactical nuclear components. (The latter can in fact include nuclear forces and weapons—such as designated numbers of Polaris or Poseidon submarines, or their missiles—also capable of use in the strategic nuclear effort.) The NATO general purpose forces are linked to, and have the backing of, the U.S. strategic nuclear force, but their targets do not include the urban industrial targets or the Soviet ICBMs, SLBMs, and other strategic nuclear forces that form the bulk of U.S. strategic force targets.

It must be stressed that the regional forces include both conventional and tactical nuclear elements, the latter essentially an augmentation of the former, not an alternative to them. The so-called conventional forces—Army and Marine divisions, Navy ships, and Air Force, Navy, and Marine aircraft—are equipped and trained to operate in a conventional-only mode or in a mode that combines conventional operations with use of tactical nuclear weapons.

99

The Opposing Forces

The opposing forces in the European area—NATO and Warsaw Pact alike—are large, powerful military forces at varying states of modernization but characteristically of generally high caliber and fighting quality. A considerable preponderance rests with the Warsaw Pact, and particularly with the Soviet Union.

Ground Forces

In Northern and Central Europe, Allied forces number some twenty-four divisions (exclusive of five French divisions, two of which are located in West Germany). Against these, twenty-seven Soviet and thirty-one other Warsaw Pact divisions are arrayed in East Germany, Poland, and Czechoslovakia, and another twenty-nine Soviet divisions immediately behind them in the western USSR. While the Warsaw Pact divisions are typically one-half to two-thirds the size of Western divisions in personnel, they are much nearer the same size in tanks. As a result, while Allied tanks total between 6,000 and 7,000 in this area, Warsaw Pact tanks assigned to their forces exceed 19,000. In the same area, Allied tactical aircraft number just over 2,000, facing a Warsaw Pact force of nearly 4,000 (many of them air defense interceptors) with a further 3,000 Soviet air defense aircraft located in the European Soviet Union.

In Southern Europe, the NATO divisions (of Italy, Greece, and Turkey) total thirty-four; they are opposed by thirty-three Warsaw Pact divisions in Bulgaria, Rumania, Hungary, and in forward positions in the southern USSR. Allied divisions in this area vary appreciably in size and state of modernization, as do certain of the Warsaw Pact divisions. In numbers of tanks, some 2,000 on the Allied side oppose 5,000 on the Warsaw Pact side. Allied tactical aircraft in this area number just under 1,000, many of them of obsolete or obsolescing types, while the Warsaw Pact inventory totals approximately 1,200, in general of appreciably newer, more capable types.

While naval strengths in general must be compared in a broader, total Atlantic/Mediterranean context, the existence of a sizable Soviet Naval Aviation force must be recognized. It is equipped with a large number of high performance strike aircraft backed by reconnaissance and other support, as well as maritime patrol and antisubmarine warfare aircraft. They have the capability of operating in strength in the Baltic and in the Barents and Norwegian Seas, with long-range reconnaissance well out into the North Atlantic. They could operate in strength in the Black Sea, but operations into the Mediterranean, lacking supportable forward bases in the Arab lands or elsewhere, would be countered and constrained by Allied air forces including those in Greece, Turkey, and Italy and the naval aircraft of the U.S. 6th Fleet.

In the category of tactical nuclear weapons, NATO has some 7,000 warheads of U.S. origin in the European area, plus significant but undisclosed numbers of British and French weapons, the former (but not the latter) covered by the established NATO control procedures—as are the U.S. weapons. The Allied weapons are deliverable by a variety of means, including land-based and carrier-based aircraft, missiles, and artillery. On the Warsaw Pact side, numbers have been estimated by NATO at 3,500, but this figure must be regarded with a considerable degree of question. The true figure may be substantially larger. The Soviet weapons are deliverable by aircraft and missiles, the latter tending overall toward longer ranges, larger warheads, but less accurate delivery than the corresponding Allied missile systems. No artillery-delivered systems have been identified in the Warsaw Pact or Soviet nuclear arsenals.

Vigorous Soviet research and development programs, followed up by large-scale reequipment and modernization efforts, have been noted for many years. On the ground, these programs have included the introduction of many thousands of T-62 tanks, with the T-72 now in production. Also, Soviet infantry has been substantially outfitted with the BMP armored infantry fighting vehicle, and Soviet towed artillery is being replaced with self-propelled artillery capable of keeping up with tanks and mechanized infantry. These force improvements, together with new surface-to-air missile systems of comparable mobility, substantially increase the capability of Warsaw Pact forces for sudden, powerful, deep-thrusting attack.

Soviet tactical aviation was substantially upgraded in effectiveness in the late 1960s and early 1970s. Increased range and added payload were first achieved, thereby expanding total attack capabilities. Improved avionics and sophisticated electronic countermeasures have also been provided. The Backfire bomber gives the Warsaw Pact a technologically advanced aircraft capable of deep penetration against Allied defenses. And new air weapons, including air-to-air and air-to-surface missiles as well as cluster-type bombs, further enhance the total potential of these forces.

At sea, additional Soviet ships are steadily entering the fleet. Characteristically, the new ships are rated as of excellent design, packed with modern arms (especially antiship missile systems) as well as communications and electronics gear, with greatest emphasis on capabilities for initial intense combat, less on sustained fighting capacity. The new Kiev-class aircraft carriers will provide the USSR with a new naval dimension—the ability to bring naval air power to far-flung areas of Soviet interest, such as Somalia, Angola, or South Asia. The Soviet submarine program continues on a diversified and large scale with the replacement of older types and steady expansion of the total inventory.

On the Allied side, a number of parallel measures of modernization and upgrading have been taken since the early 1970s, and the process continues, although unevenly from country to country and on a lower scale than the Soviet effort. Thousands of more modern tanks have been supplied to the NATO forces, particularly on the Central front, together with armored personnel carriers and highly effective Tow antitank missiles. Many hundreds of combat and maritime patrol aircraft have been added, as well as helicopters, antiaircraft guided missiles and antiaircraft guns. At sea, destroyers and other escorts, submarines, fast patrol boats, and maritime helicopters have improved the Allied naval capabilities. In general, it is clear, however, that the Soviets are closing the technological gap, where it existed, and—particularly when the large scale of their reequipment efforts is taken into account—are adding to their aggregate, available military power more rapidly than are the Allies. New NATO aircraft types—U.S. and Allied—whose planned introduction will extend well into the mid-1980s should suffice to retain a qualitative edge for the Allied forces. This will be further enhanced by precision-guided weapons ("smart bombs") entering the ammunition inventories in strength during the same period.

U.S. Interests and Policy Objectives

The interests and policy objectives that the United States looks to the NATO (including U.S.) forces to serve include the territorial integrity, political independence, and security of the member nations. The two world wars, the Truman Doctrine of 1947, and the sustained U.S. military presence in Europe attest to the validity of this interest. While complaints are periodically heard over American forces "protecting the Europeans" so long after World War II, the counterview—that it is, in the end, the United States that they are protecting there—has prevailed in every test for more than thirty years.

In an earlier chapter, we identified the main requirements that our policies of defense, deterrence, and detente impose upon our armed forces (theNATO regional forces in this case).

For defense, these requirements are:

providing as large a probability as possible of holding as much of Western Europe as possible against a Warsaw Pact attack;

bringing hostilities to a satisfactory end as quickly as possible;

keeping the loss of life and destruction of property in the area of conflict to the lowest possible level;

recognizing in all this the practical limits on what we can do if war actually occurs

For deterrence, the requirements are:

posing to the Warsaw Pact nations, and particularly to the Soviet Union, such a prospect of gains denied, as well as costs, risks, and uncertainties likely to be encountered as to dissuade them from attack, or from attempts at intimidation, interference, or coercion of NATO members based on the threat of attack

to that end, making clear that, in the event of Warsaw Pact aggression in Europe, substantial damage will be inflicted by the NATO forces on the Warsaw Pact national territories, including European Russian

restoring the deterrent even after war might have begun, by inducing the Warsaw Pact nations to halt their attack, and by dissuading them from escalation to the conflict to higher levels of destruction

For detente, the requirements are:

establishing agreed limits on the respective forces of NATO and the Warsaw Pact in the European area through the process of mutual force reductions, balanced so as to provide undiminished security

giving added stability to the military confrontation in Europe by setting limits on military exercises and movements in the European area and giving notice of such exercises and movements

establishing commitments to avoid military pressure tactics in the area

establishing broad equivalence of military strength on the two sides within the general area of force deployments in Western and Eastern Europe

The essential task objective to be served by the NATO regional forces is therefore to defend against Warsaw Pact attack against NATO member territories and forces within the NATO area by operations within the area of contact and extending into the home territories of the attacking Warsaw Pact countries.

NATO Strategy and Requirements[a]

The last comprehensive review in NATO covering strategy and requirements to fulfill such a task objective was the study entitled "Allied Defense in the Seventies"—short title, "AD-70"—conducted in 1970. It reaffirmed[b] that "NATO's approach to security in the next ten years will continue to rest on the dual concept of defense and detente," took note of increasing Soviet military

[a]Much of the material on NATO Strategy and Requirements that follows has previously appeared as an article under that title in *Survival*, International Institute for Strategic Studies (September/October 1975). I am grateful for permission to use it herein.

[b]In an official NATO information pamphlet, "Allied Defense in the Seventies," published by NATO Information Service, 1110 Brussels.

capabilities (and military budgets), endorsed the strategic concepts of "flexibility in response" and "forward defense," cited a continued requirement for an appropriate mix of nuclear and conventional forces, noted the existence of imbalances between NATO and Warsaw Pact military capabilities, and prescribed that:

In the allocation of resources, priority will be given to measures most critical to a balanced Alliance defense posture in terms of deterrent effect, ability to resist external political pressure, and the prompt availability or rapid enhancement of the forward defensive capability in a developing crisis. In addition to a capability to deter and counter major deliberate aggression, Allied forces should be so structured and organized as to be capable of dealing also with aggressions and incursions with more limited objectives associated with intimidation or the creation of *faits accomplis*, or with those aggressions which might be the result of accident or miscalculation. In short, Allied forces should be so structured and organized as to deter and counter any kind of aggression.

The statement called for attention in the following decade to important areas in NATO's conventional defense posture:

Armour/anti-armour potential; the air situation including aircraft protection; overall maritime capabilities, with special reference to anti-submarine forces; the situation on NATO's flanks; the peacetime deployment of ground forces; further improvements in Allied mobilization and reinforcement capabilities as well as in NATO communications for crisis management purposes.

Through the years following completion of AD-70, the study provided a foundation and framework for NATO's Annual Reviews of force programs. While new challenges and new problems were not lacking, no radical or general reappraisal was deemed necessary.

While it is still true that no reason for radical reassessment or revision of NATO strategy and requirements is apparent, there are by now important areas where new measures or emphases may be required.

NATO Objectives

To analyze these areas, a useful overall framework is provided by NATO's objectives. As they bear upon the operations and responsibilities of one major command—Allied Command, Europe, the collective NATO military force in the area of Western Europe, extending from the North Cape of Norway across Western Europe and the Mediterranean to the eastern borders of Turkey—four of these may. be recognized as having outstanding importance: deterrence, defense, solidarity, and detente. These have been steadfast and valid goals for many years against which to measure and evaluate NATO posture and activity; they remain so today. To them may be usefully added a fifty: equity, a fair sharing of the Alliance's burdens and risks.

The primary objective recognized within NATO circles, which NATO's military forces and strategy must serve, has been deterrence. The reason is clear. In a thermonuclear world this must certainly be the chosen means of providing security and safety for the NATO nations and people against outside threat, while safeguarding freedom, enhancing stability and confidence, and reinforcing other peaceful efforts. It is deterrence not only of major military aggression against the West, but also of temptation toward minor military operations, and toward the use of military threats and political pressures supported by military threat. It also includes means to dissuade the escalation or the prolongation of operations even after hostilities might have begun. Deterrence has worked successfully in NATO. It is interesting to note the identity in substance between this objective, as seen and served in NATO, and the policy of deterrence as we have previously developed it in the present study from the standpoint of U.S. security interests.

The second classical objective of NATO—defense—also has a dual importance. It is a strong and indispensable underpinning to deterrence, and it is the next recourse available to the NATO countries if deterrence should fail. Whereas over the years the accent within NATO circles has increasingly been placed on deterrence, the need for credible, confidence-sustaining capabilities for actual combat is no less great. It has meant that strict and thorough professional military tests and standards must be applied to, and met by, the NATO force posture and strategy.

The NATO Alliance is solely and wholly defensive in character, while the forces of the Warsaw Pact, on the other hand, are organized and deployed in a posture much more suited to offense than to defense, with armies heavy in tanks and with dispositions far forward and close to the western borders. Should aggression in Europe occur, it would only be initiated by the communist side, which would enjoy the tremendous military advantages that go with the initiative: choice of the time, the place, the mode, and the weight of attack. The advantages on the Western side are those that accrue to a defender—particularly, knowledge of the terrain and opportunity to prepare it for defense. NATO's capability for defense, despite many internal divisions and shortcomings, has proved over the years to be powerful and coordinated.

To the two classical objectives of deterrence and defense is added a third—solidarity—evidenced in the ability and determination of the Alliance to act collectively. Its values lie in denying the Soviets the opportunity to act against one nation at a time, or to play one off against another, and in the suppression of the disputes between NATO countries that have so weakened and nearly destroyed them in the past. The importance of solidarity has perhaps most powerfully been shown in the breach—specifically, in the clash of Greece and Turkey over Cyprus, and in the damage to the Alliance, and to each of them, that has ensued. On the military side, solidarity is manifested in truly collective forces (which, it must be acknowledged, NATO is still rather far from fully achieving); it is also demonstrated in unified commands and plans, common infrastructure, combined exercises, and other multinational activities and facilities. Through solidarity—working together, as General Eisenhower emphasized from the outset—the many nations achieve a level of security, which, if they acted separately, would be beyond the reach of any or of all.

A further objective to be cited is detente—reduction of international tensions. To be valid in the NATO context it should come from a reduction in the causes of tension, which means, in Europe, a reduction in the high levels of Soviet forces that are deployed in the forward areas just to the east of the Iron Curtain. For detente to have substance and reality for NATO, a reasonable degree of international stability and military equilibrium must exist between Eastern and Western Europe. In concrete terms, this means that until reductions in Soviet forces are achieved—via Mutual and Balanced Force Reductions (MBFR)—there is no basis for reducing NATO's forces. They should in fact be maintained and modernized on a scale and tempo adequate to keep a stable balance between them and Warsaw Pact forces. They should not be reduced in strength or readiness.

The fifth objective—equity—that is suggested is simply that of burden sharing. No nation should seek to gain undue advantage, or to bear an unduly low part of the total burden, at the cost of the others. The principle is easier to state than to apply, but it is evident that a sense of equity must exist if the Alliance is to remain viable. This objective, like each of the others named above, must be reflected in force requirements, posture, and strategy.

The second major determinant of NATO's needs is of course the military strength of the Warsaw Pact forces. Its principal components and trends have already been described. Within the Warsaw Pact military machine, the force structure itself in the European area has not greatly changed in recent years. Where changes have occurred has been in the degree of mechanization (through the addition of tanks and armored personnel carriers) and other technological modernization, the amount of artillery support, the improvement and expansion of forward supply bases and the complex logistical net that supports them, the increased number of aircraft devoted to the close support role, and the expansion of the Soviet Navy, with a widened scope of operations.

NATO Strategy

In the NATO strategy designed to deal with this threat—i.e., with the capabilities of the Warsaw Pact military force—the two best-known and best-recognized elements are *flexibility of response* and *forward defense*. The essence of forward defense is that NATO will not deliberately sacrifice space for time. Limited depth, and the interests of the forward countries (Norway, Denmark, West Germany, Italy, Greece, and Turkey) will not permit this. Flexible response requires defense against an aggressor's attack at a level appropriate to the time, place, and nature of the attack. As applied in NATO Europe, it could involve direct defense, deliberate escalation, or resort to all-out war.

Two further elements may be added to these, not so well recognized nor well known—collective response and undetermined duration.

Collective response is always inherent and often explicit in much that the Alliance does—in joint exercises, notably including the exercise of the standing and on-call naval forces and the Allied Command Europe Mobile Forces, both land and air; in the existence and operations of international commands such as the Allied Tactical Air Forces, the Northern and Central Army Groups of the Central Region, and the Allied Naval Forces Southern Europe (NAVSOUTH) naval command in the Mediterranean; in the multinational distribution of delivery means for tactical nuclear weapons; and in the integrated communications network, as well as other forms of common infrastructure. It is laid down in binding terms in Article V of the North Atlantic Treaty itself:

The Parties agree that an armed attack against one or more of them in Europe or North America shall be considered an attack against them all, and consequently they agree that, if such an armed attack occurs, each of them, in exercise of the right of individual or collective self-defense recognized by Article 51 of the Charter of the United Nations, will assist the Party or Parties so attacked by taking forthwith, individually and in concert with the other Parties, such action as it deems necessary, including the use of armed force, to restore and maintain the security of the North Atlantic area.

Undetermined duration is an element of NATO's strategic doctrine not so much by decision as by lack of decision. Whenever this issue has been raised, disagreement has been sharp, and the matter has been left unresolved. Some lack of clarity has been evident in the various positions that have been taken. One point of view is taken by the proponents of a "short-war" concept. They have not, however, in general, been willing to say they favored a "short-war only" posture, with its inescapable implication of readiness to accept defeat or mutual destruction should the other side continue to fight. Nor have the proponents of the other major point of view—that of outlasting the enemy—apparently been prepared to make the doctrine effective through planning for assured continuity of support. While the latter have the better of the argument, logistical shortfalls and imbalances among the countries leave the capability unprovided for. There remains a measure of ambiguity through silence on this issue.

In fact there exists an observable gap between the strategy in words of the Alliance and the strategy as expressed in facts—the facts of force strength, force readiness, force dispositions, supply levels and locations, adequate communications, and the like. For practical military reasons, absolute forward defense, on the borders themselves, is not possible—even though it is true that covering forces would engage enemy attackers as close to the borders as possible. The main defensive operations would also be conducted as far forward as possible, exploiting all available strength of the terrain. Nor can there be assurance that the integrity of NATO territory can be maintained or restored. What can be assured is that any attacking force would be subjected to heavy, continuing, and expanding losses with no certainty of tactical success, but with rapidly escalating threat to rear areas and to the aggressor's homeland. Aggressors would have to face the risk of their offensives grinding to a halt and would have to ask themselves, from a very early stage onward, just what there is west of the Iron Curtain that could justify such tremendous losses of manpower and military materiel, and such rising risk to their homelands.

Nor is the declared NATO strategy of flexibility in response fully provided for in the world of hard reality. Part of this lack comes from inflexibility within the forces themselves. Failure to standardize the equipment in NATO, past resistance to welding the separate air force contingents into true centralized commands with common systems for their employment, absence of an area logistics system that would enable ground forces to be used with adequate freedom of action, disinterest and opposition toward proposals for common procurement programs—all this takes a toll of effectiveness that may be estimated at 30 percent or more, and for some forces even 50 percent or more. There is much that could be remedied, *at lower cost*, by added initiative within NATO, and more energetic action.

There is a second, no less serious lack of flexibility, which takes the form of a curtailment of options resulting from shortfalls and shortcomings in the quantity and quality of the forces and facilities that the NATO countries, in aggregate, provide. Such deficiencies are found both in the conventional forces and in the tactical nuclear weapons that might have to be used to reinforce them. Although much has been done in recent years to expand and modernize NATO tank inventories, to mechanize forces with armored personnel carriers (APCs), to bring in Tow and comparable antitank weapons, to provide shelters for tactical aircraft, and to upgrade air defenses there are still important problems of undermanning, shortage of long-term personnel, and insufficient length of conscript training that remain to be faced and solved. The modernization of tactical nuclear weapons—specifically the 8-inch and 155mm artillery rounds as examples—reducing yields and collateral damage while gaining effectiveness through greater accuracy and improved handling under field conditions, would enlarge the range of options. These are otpions that would be available to commanders conducting the battle, and to the high political authorities, who would be seeking to control the process of escalation and bring a war to an end.

Force Structure

The structure of forces in NATO that is associated with the strategy just described comprises three main elements: conventional forces, tactical nuclear capabilities, and the strategic nuclear force.

NATO's conventional forces in Europe provide what may be termed an "intermediate" conventional capability. This is a force with a real fighting potential and significant staying power; it provides far more than a "trip-wire," yet far less than would be needed to give assured conventional-only defense in all circumstances—in other words, far less than a complete conventional capability. The trip-wire force would be a low-cost, high-risk force; it was rejected in NATO in the mid-1960s because it had lost credibility for protection against small attacks at a time when the Soviets had acquired thermonuclear weapons and the means for their delivery. A complete conventional capability, on the other hand, would be a low-risk, high-cost force; it would go far beyond what the NATO nations have shown themselves willing, through the budgets they vote and the forces they support, to maintain. Also there must be serious military doubt, in realistic terms, of the practical possibility of giving such assurance in the restricted, complex, shallow defense zones that are available in Western Europe.

The immediate consequence of settling on an "intermediate" conventional capability is a need for a tactical nuclear backup. One of the difficulties in dealing with this subject comes from the many preconceptions—and, in some cases, misconceptions—that have grown up around it. There is fear in some quarters that any significant use of tactical weapons—even the smaller weapons— would destroy Western Europe and beyond this that any use of these weapons would mean an automatic escalation to all-out nuclear exchange that would destroy the world. There are some who assert therefore that we should do away with our nuclear inventories. From another quarter comes the argument that the nuclear guarantee is the only thing that has any real meaning today—that nuclear weapons will so surely have to be used in case of attack, and are so powerful, efficient, and fearsome that there is no longer any need or place for conventional forces.

Neither side is correct. The nuclear weapon should be seen as an important— indeed indispensable—part of the tactical and strategic mix of military force in which each individual element—aircraft, tanks, soldiers, ships, and fire support— contributes to the overall defense capability, and therefore to the deterrent, in ways that are complementary and mutually reinforcing. Tactical nuclear weapons are essential because the massive added increment of firepower that they can provide helps to equalize the superior ground forces that the Warsaw Pact, exercising the initiative, could employ against us. But these nuclear weapons are like other elements of the battle that could not be used with full effectiveness in the NATO defense task except when serving as part of the larger mix.

The strategic nuclear weapons are the backup for the tactical battlefield and serve as a warning to the Warsaw Pact nations that massive retaliation is possible if they were to initiate aggression against us. These weapons are the final and the most important deterrent, but they too are not by themselves a rational or credible bar to local aggression or local intimidation.

Future Strategy and Requirements

Looking to the future, and to the probable nature of Soviet policy and dimensions of Soviet military power, it seems reasonable to assume, as regards those elements that affect NATO, that a high degree of continuity will exist in the dispositions and trends now visible. Soviet concerns over her Western borders, and her powerful interests in seeing docile rather than unfriendly regimes in control of neighboring countries, seem certain to be reflected in a continued forward deployment of large Soviet ground and tactical air forces. Above all, Soviet desire to prevent resurgent independent power in East Germany makes it imperative for her to maintain large forces there. If one accepts, as some have suggested, that the Soviet Union, although militarily strong is politically weak in Eastern Europe, then it seems almost certain that she will wish to keep an overwhelming, overawing superiority of force throughout that area; such dispositions accord well with her doctrine of mass.

Military force of this scale, so deployed, exerts a heavy influence on the Western European countries too, and manifestly deters them from pursuing policies and activities that would prove inimical to Soviet ends and interests. The Soviet Union has made amply clear her close observation of West Germany in this regard. But the line between Soviet use of her forces simply to maintain a presence and further use to conduct a more active, disruptive, and dangerous policy is one that could be easily and quickly crossed, should circumstances seem to invite them to do so. Although Soviet interests as demonstrated in the Helsinki Conference on Security and Cooperation in Europe seemed to be weighted heavily toward stability and toward achieving added legitimization for her position, her borders, and her military presence in Eastern Europe, new opportunities may bring new ambitions jeopardizing the West.

Against this background, the rationale for NATO on its part to maintain a high measure of continuity, in the period ahead, in the military posture and strategic dispositions that have been evolved through the past years, seems quite compelling. Strategically, a forward defense for Germany, Denmark, Norway, Italy, Greece, and Turkey seems to be imperative to their interests, and at the same time consistent with the interests of their allies and appropriate as a foundation-stone for Allied military planning. A stronger, more assured forward defense requires steps to overcome deficiencies where they exist in national contingents and the modernity of their forces.

Flexible response—given that the initiative at the outset of any conflict will rest with the other side—will also remain a cardinal principle of NATO strategy. In general, the range of NATO response options should be kept broad, and the conventional option should be strengthened in order to narrow the gap that now exists between the flexibility envisaged in declared policy and that which is actually provided by the forces and facilities that the nations make available. It must be reiterated that the Alliance, when it adopted the new strategy in the mid-1960s did not follow through with full and commensurate action to implement it. The prolonged delay in taking effective action to provide an integrated communication system for the use of NATO military commands is one of the foremost of the major shortcomings that could and should be remedied as early as possible.

There would be value in clarifying the issue over conflict duration, although it must be recognized that agreement will not be easy—that, in fact, full agreement may not be possible. Reluctance to contemplate prolonged destructive hostilities in Western Europe is strong, visible, and understandable. Budget-makers are highly sensitive to the costs of higher reserves of supplies as are military leaders to the competition for funds that the buildup of such supplies continually exerts. This competition will be felt in sorely needed reequipment programs, and in levels of training activity. The disenchantment with any kind of "short-war only" concept suggests, however, that NATO should lay out a carefully balanced, multiyear procurement program to achieve increased flexibility at a gradual and steady rate. Given the inherent uncertainties that would confront an aggressor contemplating hostilities, such flexibility would add usefully to the deterrent. The Warsaw Pact should be denied any basis for confidence that they could penetrate through NATO's defenses either in space or in time.

Finally, it would appear that NATO's doctrine of collective action should be reaffirmed and reinforced, and that the instances of backsliding toward a collection of forces—rather than a collective force—should be halted and reversed. The terms of commitment of forces to the NATO commands should be strengthened, and the provisions for conduct of higher training on an allied basis should be expanded. The development of allied doctrine and operational requirements should be intensified, and the authority to set and assure needed readiness standards should be tightened. Improved standardization of equipment will help to establish the necessary practical basis for such collective action. The tendencies visible in recent years by which forces of the different nations and the different services have been regressing toward more separate lives of their own should be strongly countered by the higher leadership, military and civil, of nations and of the Alliance as a whole.

The requirements that the future is likely to impose on NATO can be suggested in their essence in light of the above analysis, although their development in specific detail will inevitably be dependent upon painstaking staff work, hard decisions, and much negotiation. Time factors, in particular, will have to be carefully evaluated. The F-16 aircraft, which will replace the F-104, will be entering into operational inventories between now and the mid-1980s and must in fact be expected to serve to the end of the century, or even beyond. The choice between analog and digital techniques of communication depends greatly on planned dates of operation. And the new generations of air defense weapons, antitank weapons, artillery and tanks, and electronic warfare devices will all require careful scheduling, both for economy and to posture and train the forces to use them.

The requirement for progressive modernization of military materiel will be a vitally important one, in the future as in the past. The high levels of Soviet effort and expenditure of resources that have been sustained over many years without interruption make it clear that if NATO were to stand still it would in fact swiftly fall behind. It is not enough to develop the technologically advanced prototypes. The rate of Russian reequipment dictates a substantial, sustained reequipment program for the Allies.

As to the overall level of NATO's requirements, it may at once be said that if the Soviet Union continues to maintain and strengthen its forces, as they have done since Khrushchev's fall, there is no margin for unilateral force cuts on the NATO side. While no one can predict with accuracy the exact point, or the exact way, in which deterrence would fail, it is clear that the objective of defense—a reasonable capability in case of war of denying the conquest of Western Europe to the Soviet Union, without resorting to all-out nuclear exchange, or even in case such exchange were to occur—requires no less than now exists.

Nor do the requirements for the effectiveness of the forces—their state of training, of readiness, of logistical and other support—allow for reduction. Across the Iron Curtain, the Soviet forces in forward areas are kept continually at a level of readiness and training that would permit their quick commitment to combat. Suitable proportions of Allied forces must be kept at properly related states of readiness on our side.

There has been interest in several quarters[c] during recent years in the feasibility and desirability of structuring Allied forces to put top priority on those elements that would be needed earliest in case of hostilities. With this principle itself there is no quarrel. Some of this work goes on to challenge existing structure severely on these grounds, and to offer a range of proposals, many of them interesting and innovative, to accomplish such restructuring. Obviously, such proposals require thorough and searching analysis, and a careful weighing against many hostile courses of action, before they can be considered for adoption, but there is value nonetheless in the challenge that has been made. On the other hand, recalling the repeated surveys and studies that have been

[c]See, for example, Steven Canby, *The Alliance and Europe: Part IV, Military Doctrine and Technology*, Adelphi Paper No. 109, The International Institute for Strategic Studies (London, 1975).

made by commanders in Europe to squeeze out redundant, outdated, and lower priority elements, and to find resources to fill new needs, it may be anticipated that the changes will prove to be less than claimed, although the proper test of the matter can only be carried out through careful detailed study.

The more serious hazard is that the thesis will be stretched to assert that the kinds of forces that would be used to sustain combat for a longer period are not really needed. In the past, inklings of this argument have been heard in Congressional circles in the United States. For reasons mentioned in the discussion of the "short-war only" issue, such a proposition does not appear to accord with NATO's deterrent and defense needs.

The only way now apparent in which NATO forces could safely be reduced is through mutual force reductions—the goal of the MBFR negotiations that began in Vienna in 1973, aimed at reducing NATO and Warsaw Pact military forces in the central European area. Perhaps at a later time the process could be extended to the Southern/Mediterranean and the Northern sectors as well. If carried out on the prescribed principle of "undiminished security," sizable cuts could be made on both sides, and added stability could be the result. At what point the process might reverse and the situation become less stable—because practical ties among Allied forces had been weakened, for example, or because a kind of vacuum had been created through insufficiency of NATO forces in relation to the breadth of the area for which they are responsible—it is very hard to say. The NATO countries, between the time the idea of MBFR moved from rhetorical proposal to imminent negotiation, i.e., between 1968 and 1973, adopted a much more cautious and apprehensive approach. The NATO nations in the Mediterranean have shown no desire to apply MBFR in that area, and the Scandinavian forces in the North are in any case so small in relation to Warsaw Pact forces in the Baltic area, and in the Kola Peninsula, as to offer little opportunity for further reduction. It may well be in any case that the practical limit will in fact be set from the other side—that is, from Soviet consideration of the minimum force that *they* wish to maintain on the territories of their Eastern European allies, over whom they will certainly wish to preserve an overwhelming military superiority.

As to what reductions would provide for undiminished security, careful note must be taken of three relative NATO disadvantages, or asymmetries, that work in favor of the Warsaw Pact. As mentioned earlier, the military initiative must be accorded to the communist side in our calculations, since NATO, in case of aggression, cannot be other than the defending party. The NATO disadvantage on this count is aggravated by a less favorable geographical situation—wide, shallow defense zones broken by water barriers and neutral nations, and distance differentials such that forces taken back to the Soviet Union would be only one-eighth as far from the forward area as those taken back to the United States; moreover, the latter would have an ocean passage by air and sea to manage in order to return. And finally, as earlier noted, the Soviet forces are so constituted as to put major emphasis on the kinds of components that have offensive capabilities—tanks, armored personnel carriers, and now self-propelled artillery, as well as large, modern tactical air armies. The Allied forces, on the other hand, show a more balanced composition.

If equal reductions, or even equal percentage reductions, were applied to the two sides, the result would be forces even more unequal than they are now in their respective capabilities. The asymmetrical reductions that have been sought by NATO are therefore a necessity if undiminished security is to be preserved. One proposal—for "common ceilings" for the two sides at an appreciably lower level than now exists—has much to commend it. The seriousness of Soviet interest in MBFR could not be fully determined in the early negotiations. It may be revealed in her reaction to proposals of this kind. But pending suitable agreement on *mutual* reductions, the strength of the NATO forces must be maintained. The Soviets would have little reason to negotiate reductions in their forces if they can expect the West to reduce unilaterally.

Beyond the reasons of deterrence, defense, and detente, there are considerations of solidarity and equity that weigh against any significant unilateral cuts either in U.S. forces or in European forces.

As to the American forces, it has been a well-established European view, frequently given heartfelt expression, that European forces could not and would not make up the deficit that a unilateral U.S. withdrawal not matched by a Soviet reduction would inevitably cause. The presence of U.S. forces, in General Eisenhower's phrase, "put the glue into NATO" by making a tangible and visible American commitment in the early 1950s that led others to do the same. They also put the starch into NATO—the necessary confidence that a defense posture could and would be developed that would deter further Soviet advance. Some of the events of more recent years—the questions by General de Gaulle, for example, whether any country but one's own (the U.S. included) would risk its own nuclear destruction for an ally, and the American abandonment of its allies in Southeast Asia—have brought a measure of doubt about the American commitment that adds to the need for a special effort and steadfastness if it is to be contained and overcome. The picture of unilateral American force withdrawal from Europe, no matter how it might be rationalized, would be bound to feed the doubts. The problem is heightened because such withdrawal would leave important areas visibly weakened where existing forces already are known to be militarily no more than adequate, if that.

As to any unilateral European reduction, the strong current of opinion that already exists in the United States that the allies are doing less than their share would be sharply aggravated. The fact that this view is in large part a misconception, at least in regard to certain allies and probably overall, would be unlikely to change the reaction. The fact is that the United States provides no more than 15 to 25 percent of the total army, navy, and air forces in NATO Europe. And it is true as well that the U.S. forces in Europe make a far greater contribution to U.S. security there than they would if they were located in the United States. But it is also clear that the U.S. effort exceeds that of its allies by nearly any measure that can be devised.[d] Since the late 1950s, a similar concern has existed in the United States over the adverse balance of payments impact that the United States sustains by maintaining its military forces in Europe. In part this concern has been alleviated by the offsetting effect of European purchases of military equipment from the United States, but the negative attitudes persist and are widely held.

On either of these counts—budgetary effort and balance of payments effect—any significant step by the Europeans in a direction disadvantageous to the United States, specifically through reductions of their defense forces or their defense budgets, could be expected to have critically damaging effects in the United States.

Future Requirements of NATO

Among the important requirements of NATO in the future, a greater standardization of arms and equipment has received repeated recognition but little action. In important areas, the actual trend in past years has, in reality, been toward less standardization rather than more. In the early 1970s, the types of tactical aircraft in use in NATO had reached a total of twenty-three—and the figure would have been thirty-nine if significant variations of models had also been included. An exercise of the Allied Command Europe Mobile Force (Air)—a composite, quick-reaction force designed to deploy on short notice to a threatened area to present "many flags" to a possible aggressor—saw five national fighter squadrons assemble, each with a different type of aircraft, supported by its own separate air line of communications running back to its own home air depots, unable to use the bombs, wing-tanks, and ammunition of the host country in which it was deployed. It was an achievement in NATO that four countries adopted the same gun for their mine-sweepers, but it was soon reported that each had made enough national modifications to require the use of

dIn the early 1970s, when the total American defense budget was running at $80 billion, the combined budgets of all other NATO countries were just over one-half as great. In terms of defense expenditures as a percentage of GNP, when the U.S. figure was over 6 percent, the figure for the rest of NATO was below 4 percent. The American figure in fact exceeded the figure for every other country. In terms of defense expenditure per capita (a measurement favored by some German authorities, in particular) the U.S. outlay was slightly under $400 a year while the rest of NATO ran less than one-third as much. Only in one series—compiled by the author in an effort to reflect greater equity—was the picture more balanced. If, for each country defense expenditures as a percentage of GNP are correlated with per capita income (having in mind that countries with higher per capita income could reasonably be expected to bear a higher percentage burden) the U.S. and NATO as a whole came out very nearly the same. It must be noted however that the figure for "NATO, other than the U.S." is strongly and favorably biased by the inclusion of Turkey's large population and relatively low GNP.

different ammunition. Such instances are legion. The cost in lost effectiveness is obvious and staggering; the "opportunity cost" in additional, more modern materiel that could otherwise be acquired is equally impressive. Above all, this has been a failure of leadership, with important overall needs subordinated to the separate interests of military bureaucracies, commercial special interests, and their allies in governments and parliaments. The pressures for positive action are building up, however, as budgets feel the double squeeze of financial stringency and escalating prices—and as the spectacle of wastefulness becomes more widely known. Initiative on a European scale—the commitment of European Defense Ministers to consult with each other before adopting a new major weapon system, for example—offers a promising beginning. Some interest has also been displayed by the U.S. Secretary of Defense and the Armed Services Committee of the U.S. Senate in the same direction. In addition, serious analyses are beginning to appear offering ways to overcome in an equitable way the potentially prejudicial economic and technological effects, on particular producers and particular nations, that increased standardization might involve. It is to be hoped that these stirrings will gain adherents and increased momentum.

Special issues and requirements exist within NATO in the area of tactical nuclear weapons; here, a number of practical measures of improvement are possible and necessary. Planning work already well begun needs to be vigorously pursued to develop and refine selective employment plans for these weapons— plans tailored, for example, to "battlefield" use in particular areas of likely threat—in order to provide well thought out options, preplanned or partially preplanned as appropriate. Such options make a special contribution to flexibility of choice and control of escalation by the high political leadership of the alliance. It is of note that a much more positive attitude toward such options has been emerging in Western Europe.

Improved small-yield weapons should be developed. The 8" and 155mm nuclear artillery weapons are of particular importance. Greater accuracy, reduced collateral effects, and greater field-readiness are among the characteristics to be pursued. The change here is not of radical or conceptual nature; weapons of these sizes already exist, as do procedures for their controlled employment. The scare-stories that appeared about "mini-nukes" (suggesting that they connote a loss of political control over nuclear plans and operations) were quite unfounded. In reality these improved weapons will add to the flexibility of choice and thus strengthen the control-in-fact that political leaders can exercise over the developing situation, in case the use of tactical nuclear weapons in combat in Western Europe should ever become necessary.

Continued work in the area of limiting the collateral damage that nuclear weapons would create is also required. This is a matter of the most serious concern to SACEUR and to other NATO military leaders at all levels. Here a special degree of precision and care is required. Wide misconceptions exist not only within the informed public, but also within responsible governmental circles as well, which are only gradually and partially being removed. It appears that many of these grew out of early studies and publicity, particularly in the United States and Britain, largely through errors in interpretation, extrapolation, or the aggregation of highly constrained uses with others having vastly greater effects. A sober approach to the use of these necessary but dangerous weapons is required. Not only the design of weapons but the selection of which one to use and the tactics, locations, and conditions for using them are important. By such means, collateral losses and damage can be reduced by factors of a hundred or more below what would otherwise occur.

The streamlining of procedures to improve timeliness and effectiveness of nuclear weapons use, without impairment of essential control, should also be pressed. Alternative means are needed for providing the voluminous intelligence and operational information necessary to policy decision; practical techniques for such information handling are available, and their development should be accelerated. Selective use planning will assist further. A new impetus needs to be given to the NATO Integrated Communication System (NICS) program, too long stalled by top-level NATO inertia and national obstruction.

There is a further, wider range of requirements that bear strongly on NATO's ability to maintain the military strength that will counterbalance the military might of the Warsaw Pact nations. They include the need to safeguard the access of NATO countries to Middle Eastern oil; to deal with divisive economic problems; to repair breaches such as that between Turkey and Greece over Cyprus; and to find better ways of conveying a comprehensible and adequate picture of NATO military activities and needs to publics and parliaments. These broader issues have come to represent a particularly severe challenge to NATO, if only because NATO has shown itself thus far to be ill-equipped to deal with them effectively.

A long agenda of requirements inevitably reflecting some sense of inadequacy or shortcoming, at least in detail, risks leaving an impression of failure or infeasibility. In the case of NATO, the true situation, despite the real problems that have been suggested, is just the opposite. By any relevant set of objectives, an alliance that has contributed in a major way—as NATO has—to peace in Europe for a quarter-century, without loss of territory or of freedom, must be judged a tremendous success. Its resources of wealth and talent are such that, if attention is given to the requirements that confront it—a crucially important "if"—its success can certainly continue.

Implications for U.S. Force Programs

The requirements that fall on NATO fall importantly on the United States, as its strongest member and pace setter.

A first principle for NATO is to maintain the military force balance with the

Warsaw Pact forces. Until Warsaw Pact, and specifically Soviet forces are reduced, no basis in security exists for reduction of American forces, assuming that Western European countries would not be able nor willing to pick up the slack. If the European countries should make tangible progress toward a European Community—in ways that would make of themselves a single effective center of administrative, technological, and military power and decision—then a greater assumption of the military burden by them would obviously be in order. The likelihood that such progress will be forthcoming in the near future does not appear great, although the steps that are being taken by the Eurogroup are constructive, encouraging, and contributory to a stronger spirit of European purpose and identity.

The second NATO need, which also poses major requirements for the United States, is to maintain the qualitative balance with the Warsaw Pact countries in modern military technology, as incorporated in the armed forces. In recent years, the Soviets have been rapidly closing the technological gap—to the point where it is a mistake to believe that a Western technological superiority exists that could substantially offset numerical inferiority. Soviet submarines are noisier (although of excellent performance, and there are many of them), but their surface ships are to top quality, especially in their missile and gun armament and electronics. Although their air forces lagged for a time in range and load-carrying capacity for close air support, newer model aircraft are greatly improved in these categories and are entering the operational inventory in substantial numbers. On the ground, their tanks, personnel carriers, and artillery—especially the BMP carrier and the new self-propelled artillery—measure up well against anything the NATO forces have. They excel older equipment-types still found in sizable quantities in many NATO armed forces.

It is a strength of NATO that its attack on this problem is so diversified. It is by no means the case that the United States alone bears the burden of innovation and invention. Over the years Britain, France, Germany, and Italy have contributed major advancements in a wide range of weapons and equipment types, as have Belgium, the Netherlands, and Norway in particular sectors such as small arms, electronics, and ship-borne missiles. NATO's weakness is that to date there has been a too-general failure collectively to select and standardize on particular items for new generations of weapons and equipment. The need to do this is becoming steadily more imperative, and the evidence of a recognition of the need is becoming stronger within the NATO ministries of defense— notably including the U.S. Department of Defense as well as elements within the U.S. Congress.

The third principal need or goal for NATO, in which the U.S. has properly joined, is that of mutual force reductions in Europe. In the first stages such reductions are not likely to offer significant money savings. In fact, for the United States, some analyses indicate that the net result might be some added costs, at least initially. If U.S. units withdrawn from Europe are maintained in being at high readiness in the United States (as may well be necessary if the Soviets simply pull their forces back to join other units in their Western Military Districts), several outlays will be required. Land may need to be acquired at possible substantial cost; it is provided free of charge in West Germany today. New barracks may have to be built, additional sets of training equipment provided (leaving the combat sets stocked and maintained in Europe), and an added increment of airlift capacity for the rapid return of withdrawn units (ground, air, or both) provided.

Nevertheless, the other benefits of mutual force reduction may outweigh the monetary implications. If the reduction on the Soviet side is concentrated on tank-heavy units, the cut-backs should have a stabilizing effect. The offensive weighting of the Soviet force composition will be diminished. In addition, smaller forces will mean that the relationship of the total Warsaw Pact force to the area they would have to overrun to achieve a quick victory will be somewhat decreased, and thus improved from a Western standpoint. (The relationship of reduced Western forces to the area they would have to defend would also be lowered, but the net effect should be somewhat greater in relation to the Soviet forces than to those of the West.)

One further point needs to be made before considering the specific forces and program activities the United States should support in behalf of the NATO commitment. NATO is not the only potential arena of need for employment of American "general purpose" forces, but it is by all odds the most important and compelling one. The composition, weaponry, and support of such forces, and their resulting military capabilities, should therefore be strongly weighted toward NATO needs and toward the conditions of potential combat in the Western European, Mediterranean, and North Atlantic areas. While the requirements of other areas as well as the internal needs of the U.S. armed services for their own stability and long-term effectiveness should not be neglected, standardization of equipment and development of doctrine and organizational structure best suited to the support of the NATO commitment are worthy of greater attention than they have habitually received in the past.

Army Forces and Programs

Foremost among Army needs is that of maintaining the present level of forces in Europe—four full divisions plus a brigade from each of three others. An additional brigade is located in Berlin. Backing up these forces are ten Army divisions and two Marine divisions in the United States. Unless and until Warsaw Pact forces in Europe are reduced, the need for the American forces must be expected to continue. In particular, those in Europe will need to be kept there.

An important Army training exercise is the REFORGER program, by which a force of division-minus size is deployed by air from the United States, "married up" with its stockpiled equipment in Germany, and put through field exercises to test its readiness and give operating experience in the NATO European environment. Although expensive, the exercise is of signal value in translating concepts and plans for rapid reinforcement of the European theater into practical reality. Where initially (in the late 1960s) the same division was repeatedly deployed—the First Division at Fort Riley, Kansas, which had recently been brought back from Europe, except for one of its brigades—on occasion in recent years other divisions have been utilized, thus spreading the training and experience more widely. It is a program of sufficient value to true operational effectiveness to warrant its continuation.

Periodically suggestions are made that the sectors of the four American divisions in Europe—now generally located in the Frankfurt-Nuremberg area—should be shifted to positions further north, closer to the main axis of threat across the North German plain. For a variety of reasons, the change has never been made, or formally proposed. The expense of relocating the units and their supporting installations (and the difficulty of acquiring the additional land that would be needed) would be very great. Also the approaches toward Frankfurt, in particular, are themselves of key strategic importance to a Western European defense. And finally, the possibility exists that the powerful American forces, on the flank of Warsaw Pact forces attacking westward in north Germany, would be well placed to cut off a major drive in the north, particularly if French forces were participating in the battle and could join in the counteroffensive. Finally, there is a logic to placing the Netherlands, British, and Belgian corps in positions that guard the approaches to their own homelands.

Major steps in the modernization process involving the United States are in prospect, and they are needed. A new tank to replace the U.S. M-60 is high on the list. In European armies, a successor to the German Leopard tank is needed. Both the U.S. and the German forces are designing a new tank. It is to be hoped that a single type will be chosen, or, if that effort fails, that the same gun, engine, power-train, suspension, and tracks will be used on all.

A new mechanized infantry combat vehicle (MICV) is needed and planned to replace the armored personnel carrier that is now the mainstay of NATO mechanized divisions. The MICV, which will allow infantrymen to fight from the vehicle, will add to the mobility of operations, and provide a closer match for the Soviet BMP, also a fighting vehicle, which is entering the Warsaw Pact forces in large numbers.

The density of antitank missile systems in U.S. and Allied ground forces merits further increases. On the basis of experience both in the Middle East war and the Vietnam conflict (experience on both sides, in both cases) the upper limit for the appropriate mix of such weapons has not yet been reached. The Tow (vehicle-mounted) and Dragon (man-portable) systems are of high effectiveness. A Tow-equipped attack helicopter promises to be highly effective and cost-effective.

Improved air defense weapons are needed. The European Roland II, scheduled for adoption by the U.S. Army as its short-range all-weather air defense system, will help fill a weak area long existing in present armament and will serve as a contribution to weapons standardization. SAM-D, an improved medium- and high-altitude air defense system will introduce a new generation, which may provide a replacement for the Hawk and Nike systems in future years.

A wide range of additional technological advances are under development or under study, extending from new artillery and new ammunition to more effective surveillance, target acquisition, and fire control systems. While diverse and numerous, they contribute to defense by providing added firepower capabilities, and help to maintain a favorable qualitative balance in relation to Warsaw Pact forces.

A number of the U.S. divisions located in the United States are "light" divisions—i.e., infantry, airborne, or air assault rather than armored or mechanized. For NATO employment, the armored and mechanized units have greater power and mobility against Soviet-style forces. Actions announced in early 1976 to bring two more of the U.S. Army's divisions to "heavy" status reflect a welcome consideration of the NATO needs as the principal basis for Army force planning. The provision of additional antitank weapons to all Army and Marine divisions likewise gives them added capabilities for service in the NATO environment.

Proper logistic support is a major and continuing concern for Army forces in Europe, or destined for Europe, in case of hostilities. European stocks have been drawn down on important occasions in the past—during the Vietnam and Arab-Israeli wars, for example—and they have been slow in being rebuilt. So long as major forces are planned to operate in Europe, such draw-downs are to be expected. Their duration should be kept short, however, by the maintenance of U.S. stand-by and "surge" capacity to expand output rapidly in such items as tanks, armored personnel carriers (APCs), helicopters, and antitank and air-defense missile systems.

Navy Forces and Programs

Consideration of U.S. naval needs in the NATO context will be somewhat abbreviated at this point, since more extensive naval problems will be examined in a later section of this study, and since, in any case, many naval issues involve special technical questions best left to the sailors to develop in detail.

The outstanding and continuing U.S. naval issue in the European NATO area is the maintenance of two carriers in the Mediterranean. The issue becomes increasingly acute with Soviet improvement in its attack submarine fleet, establishment and strengthening of its sizable Mediterranean squadron, and addition of the Backfire bomber to its already potent naval aviation. Obviously, the carrier task forces operate in an environment of some threat, worsened by turmoil and instability in the region, including the effects of the clash of Turkey and Greece over Cyprus, the uncertainties concerning the direction future Italian governments may take, the Palestinian issue and the civil strife in Lebanon, the evolution underway in Spain and Portugal, and the background of concern over future events in Yugoslavia.

While such considerations as these add concern regarding the presence of the Sixth Fleet, they add even more to the need for its presence. It is a force that concurrently serves multiple purposes—from maintaining valuable lines of contact and contributing to stability and allied confidence in peacetime, to providing a capability for quick, flexible reinforcement and support in time of crisis or conflict, together with a powerful capability for helping to clear the Mediterranean of hostile naval forces, besides establishing naval control of the area if hostilities should occur. Even if risks rise in the Eastern Mediterranean basin due to developments in Greece, Turkey, Italy and Yugoslavia, the flexibility of the force should permit its presence to continue on a steady or intermittent basis and the stabilizing effect of its presence to continue to be exerted.

It is also necessary to remember that the carrier forces are not alone. Friendly submarines as well as other surface ships are operating in the same area, and large Allied land-based tactical air forces (including U.S. components) are available and are coordinated in their command and control with the operations of the Sixth Fleet. While naval commanders may have on occasion tended to forget about these other friendly forces, and to feel alone and vulnerable "out there," we may be confident that the Soviet admirals have not.

Use of the full flexibility of the carrier forces would permit the exit of one or both of them from the Mediterranean from time to time to conduct exercises nearby. This has been done occasionally in the past. It could be done more often if one could be confident that the practice would not become a cover for gradually withdrawing them. European concern and skepticism on this issue are an added factor to be recognized.

The U.S. Marine contingent in the Mediterranean, normally of battalion size, provides an additional highly flexible contribution. It is most effective when combined with its helicopter movement capability. When helicopters have been absent on occasions in the past, the Marine capability has been drastically reduced in terms of response times; the helicopters provide useful emergency lift in crisis and disaster situations. United States planning for the European area should include the assured presence of this helicopter lift at all times.

While U.S. naval forces in the Mediterranean form a highly self-contained force, the naval bases from which they operate contribute a great deal to their effectiveness. Training, maintenance, rest, and replenishment are among the functions served. Stringent restrictions on use of oil have added to in-port times in recent years. Without such basing and home-porting arrangements, the retention of the American naval forces in the area on a sustained basis at anything close to the levels maintained in the past would become impracticable.

Several of the Navy's equipment and development programs are of key significance to the U.S. military position in the European area. Provision of additional modern escorts can enhance the security of the carrier forces, and additional attack submarines should assure the ability to maintain planned submarine levels in the area. The high-performance F-14 aircraft provides the carriers with added capabilities to achieve and maintain air superiority in their operating areas, while the Harpoon provides a long-needed antiship missile system. The Aegis area air defense system is designed to deal with the current threat and provide growth potential for the future.

NATO has lagged badly in the past in providing for effective standards of coordination among the naval forces deployed in the Mediterranean. In addition to the higher command structure, a few useful steps have been taken in the establishment of surveillance centers and the activation from time to time of the "Naval On-Call Force, Mediterranean"—a multinational force that has worked out in its exercises improved techniques of naval cooperation. The demanding mission of the American carrier task forces imposes special requirements of communications and data-display, but beyond these there are practical measures of Allied interoperability and mutual support that could be developed and introduced into the operating forces.

Air Forces and Air Force Programs

The American tactical air contribution to NATO holds a key importance in the overall NATO effort for several reasons. First, it is substantial in size; the 500-plus total U.S. first-line combat aircraft in Europe roughly match the respective total air orders of battle of the United Kingdom, France, and West Germany, each possessing about 500 aircraft of various types. The twenty U.S. fighter squadrons in Europe are backed up by four additional squadrons assigned to Europe but "dual-based" in the United States, and kept at a high state of readiness for deployment to prepared bases in Europe. And behind these forces stand more than a thousand additional tactical aircraft, comparably modern in overall composition, most of which could and would be quickly deployed to Europe in case of major hostilities there.

Second, the American force is a pace setter for the whole NATO air effort. Its aircraft—F-4 and F-111, in recent years, soon to include the F-15, F-16 and A-10—represent the most advanced technology in existence. Manned by crews unexcelled in training and technique, they provide qualitative superiority over the Warsaw Pact forces. And the concepts of employment, and of command and control built into these forces provide a doctrinal model of great value for modern air operations.

Finally, the air capability of which they are part occupies a key place in the NATO strategy. The battle for air superiority should give the Western air forces increased freedom of action to attack Warsaw Pact military targets—their attacking forces and supporting activities, in particular—and at the same time do much to keep Warsaw Pact air "off the backs" of NATO forces and installations. In their second major role—close air support for the NATO ground forces— NATO tactical air offers a crucially important means of responding, with massive concentration and quick flexibility, against the main axes of Warsaw Pact attack when and where it develops, and particularly where it threatens a major breakthrough. Their third role is interdiction—deep or shallow attacks against fixed facilities and installations essential to the support and continued momentum of the communist attack.

There are significant issues and disagreements, and important shortcomings, regarding the capabilities and concepts for the employment of these forces, but even taking these into full account, the NATO air forces provide a uniquely powerful and flexible means for NATO commanders to shape the battle in case of Communist aggression.

One of the major issues has been the difficulty of establishing a single concept and system for the employment of air resources in each of NATO's principal regions (North, Center, and South/Mediterranean). It is critically important that air resources of all national contingents be capable of being directed and applied, in the close air support role, to assist the ground forces of any particular nation needing such support. The techniques for providing such support effectively and in time are complex and demanding in the training, the communications facilities, the radars and other installations they require. It is difficult enough to master a single system. Where systems differ, the support will be feeble at best, or nonexistent. In the Central Region, deep-rooted differences of concept and doctrine—for example, in the relative weight given to close air support versus interdiction—have long hampered and detracted from full air effectiveness and flexibility. Such differences are reflected in aircraft designed for differing roles, in differing armament and electronics, and in contrasting systems of flight planning and in-flight diversion. Such differences not only contribute to a lack of standardization and cross-servicing capabilities, but are in turn reinforced and aggravated by such deficiencies.

Steps to alleviate these conditions are one of NATO's major military and political responsibilities. Such steps, reversing prior trends toward separatism, have been underway—particularly since the early 1970s—but the task is large and difficult in relation to the rate of progress made. The continuing resolution of these problems, and the *adaptation of American tactical air forces* (together with other NATO tactical air forces) *to a common system of employment* remains a need of the highest priority.

If aircraft costs were not so high, or defense budgets not so stringent, it would be desirable to outfit NATO air forces—the United States included—with aircraft of nothing but the highest levels of operational quality and performance. As it is, some compromise between quality and numbers has had to be reached. A NATO force composed solely of the highest performance aircraft (at costs in the order of $20 million each) would necessarily be so small that against the large Warsaw Pact air order of battle, in the extended geographical area from Norway to Turkey, it would be utterly inadequate. The solution adopted by the United States—the establishment of a "high-low mix"—poses a difficult task of analyzing the respective numbers of specific high-performance and lower-performance aircraft needed to deal with the specific Warsaw Pact air order of battle existing and foreseen. The proper employment of such a combination of aircraft will impose new requirements on the generalship and decisionmaking ability of air commanders and their staffs.

A new generation of aircraft apparently highly suitable for these purposes is now at or near operational readiness. The U.S. F-15 is widely regarded as the most capable land-based fighter in the world at this time; it takes its place at the top of the high-low mix. The F-16, a lighter-weight fighter adopted by the United States as well as a four-nation NATO consortium (Belgium, Denmark, the Netherlands, and Norway) is viewed as the lower-cost complement to the more sophisticated F-15. It will provide certain multipurpose capabilities, i.e., a ground attack capability in addition to an air superiority. The A-10 is being developed as a close support aircraft that will be effective against the large Warsaw Pact armor threat. The F-111 will continue in its interdiction role.

The ground attack capabilities of these aircraft (and of the F-4s now in the theater) are being greatly enhanced by the introduction of precision-guided munitions (PGMs), the "smart bombs" demonstrated in Vietnam. These seem certain to multiply the effectiveness of the aircraft inventory, and provide a return commensurate with the costs and capabilities of these advanced aircraft. What the limiting effects of European weather on laser and television technology in practice will be—in the frequent fog and overcast that is characteristic of much of the year—will take time and experience to determine, as will the practicability of providing radar-guided versions to overcome this difficulty without pricing the weapons out of a competitive range. More traditional weapons—including bomblets giving area coverage—delivered with greatly improved navigational accuracy seem likely to retain their importance along with the PGMs.

Such improvements in accurate navigation—including the long-awaited LORAN-D system that was well-proved in Vietnam—should enhance the effectiveness of conventional close air support and permit more accurate delivery of smaller-yield nuclear weapons, thus reducing the collateral damage resulting from such use.

A further major step in enhancing the effectiveness of tactical air power in Europe can come from the introduction of the AWACS (air warning and control system). A first priority requirement for this system in Europe is as an airborne radar platform capable of tracking hostile aircraft in the air, whatever their altitude, at distances of a hundred miles or more. For NATO, it offers an unexcelled means of "gap-filling" against low-level aircraft underflying the large, fixed NATO radars, and of backing up these radars, which would be highly vulnerable to early destruction in case of war. Provision of additional capabilities enabling these aircraft to be used for airborne control of tactical air operations will require careful NATO study. Such use is a promising possibility that must be weighed against the concepts and organizations that are finally established for integrated air operations in the principal regions. The AWACS is not free of problems—its cost, its vulnerability to attack by enemy air, and its susceptibility to interference through jamming—but these are of the type to be dealt with through the design of the system and the procedures under which the impressive new capability will be used.

A word may be said about the state of electronics countermeasures of planning and preparations in the tactical air forces of NATO in general and the United States in particular. Experience has shown that the suppression of enemy air defenses that is essential to effective close air support at tolerable costs is greatly dependent on ECM. Yet progress has been halting and fragmentary. It is an arena of move and countermove, in which a "closed loop" is required that will detect an enemy's latest innovation, design and develop a hardware or software counter to it, speed it to the field forces, and incorporate it in their equipment and procedures. United States initiative, and assistance to the NATO air forces, will be a continuing need.

The provision of shelters to protect aircraft on the ground from quick and easy destruction by opposing air forces has been one of NATO's major achievements in recent years. The six-day Arab-Israeli war in 1967 gave conclusive proof of the operational need for such shelters. In addition, by reducing vulnerability to surprise attack, they contribute stability to a military confrontation situation. Conservative studies have indicated that, in conventional hostilities, aircraft protected on the ground by shelters could be expected to survive to carry out ten times or more close air support than they could, had they been exposed, unsheltered, to hostile air attack. The United States has been a leader in NATO in advancing and supporting this program. It can be profitably extended to include all NATO combat aircraft in Europe, plus those firmly committed for immediate deployment. The U.S. Defense Department has supported this program constructively, and its support is well warranted.

A final requirement, sometimes neglected, is provision for logistic support of U.S. air forces assigned to NATO. Ammunition and fuel (POL) are of preeminent importance. As in the case of the Army and the Navy, ammunition stocks have in the past been drawn down for such demands as the Vietnam war and the provision of arms to Israel. What is required is a more flexible system to restore the deficits quickly, and an adequate budgetary priority for doing so. The POL situation is more complicated. Since the French government withdrew its forces from the NATO command in 1967 and required the removal of NATO forces from French territory, France has continued to operate the pipelines across France supplying U.S. air forces in Germany, but without assurance that such operations would continue in war. France has stated it would only decide at that time whether or not it would stand with its NATO Allies. Although means exist to restructure the pipeline net at moderate cost to permit supply through the Low Countries, NATO, and in particular the United States, have been laggard and resistant toward doing so. The consequence has been to leave open the possibility of having to rely on improvised, overtaxed and vulnerable truck, rail, and barge transport. If inadequate, they would inevitably result in failure to obtain the full firepower contribution that air forces in Europe could make and delay in moving U.S. air reinforcements to the critical Central Region.

Tactical Nuclear Weapons and Programs

The United States contributes to the aggregate tactical nuclear capability in several essential ways. First is the maintenance of U.S. nuclear delivery systems together with the nuclear weapons they would utilize. This is a substantial and diversified inventory of Army, Navy, and Air Force weapons and delivery systems. Second is a series of programs of cooperation between the United States and most of its NATO Allies by which the participating allies maintain delivery systems for U.S. weapons, and the United States provides (under U.S. custody) the weapons they would utilize. The allies are thus enabled to take an active, intimate, and informed part in nuclear planning and operations that would vitally affect their defense and safety in time of war. And third, the United States has furnished nuclear technology and components to the United Kingdom, which has committed its nuclear weapons to NATO subject to circumstances of supreme national need.

The stockpile of American nuclear weapons in Europe for the support of both U.S. and Allied delivery systems was established at the level of some 7,000 weapons more than a decade ago. The range of yields is wide—from subkiloton at the lower end to more than a hundred kilotons at the high end of the spectrum. The weapons and their delivery systems include missile and tube artillery for use against battlefield and battlefield support targets (Lance, Honest John, and 155mm and 8-inch artillery); tactical aircraft of several types; Pershing surface-to-surface missiles; Polaris and Poseidon submarine weapons for use against military targets throughout the theater; and special purpose systems including air-defense systems on land and at sea, antisubmarine warfare weapons, and atomic demolition munitions (ADMs).

Standards of readiness and maintenance are high for all these weapons, and standards of security extremely high. Significant measures of modernization and improvement are underway, providing tailored effects, greater accuracy (permitting reduced yield), lowered collateral effects, and further heightened security.

While the number 7,000 has no particular magic to it, it nevertheless represents a considered evaluation of NATO needs. Overall, the relation between the existing inventory and the total array of significant targets that Warsaw Pact forces and their supporting installations and activities (logistical, communications, and the like) would present in case of war has been carefully examined and repeatedly reviewed. A large number of the small-yield battlefield-type weapons need to be deployed widely across the front because of the short range of their delivery systems and the possible requirement for their use on such short notice as to preclude long-range transport. Air defense and antisubmarine weapons as well as ADMs are subject to some of the same difficulties. Air-delivered weapons and submarine-launched missiles offer greater range of coverage.

As modernization proceeds, conflicting trends will affect the numbers needed. Increased range of delivery means (the shift from Honest John and Sergeant to the longer-range Lance, for example) tends to reduce total requirements. At the same time, the adoption of more accurate smaller-yield weapons— if several of them are to substitute for one larger one in order to reduce collateral damage—may increase numbers while reducing total yield. (It might be noted in this regard that the aggregate yield of several thousand of the small battlefield weapons—in the kiloton range—does not equal the yield of a single Soviet SS-9 strategic nuclear missile at twenty-five megatons.)

Among the steps of nuclear modernization being taken by the United States are the provision of the B-61 improved aerial bomb, the introduction of the Lance surface-to-surface missile, and the development of a terminally guided Pershing warhead that offers extreme accuracy, lower required yield, and reduced collateral damage. There is need as well to proceed with development of a longer-range, improved 155mm nuclear artillery round.

All NATO nations share a strong interest in continued steps by the United States (in cooperation with NATO host nations) to tighten and strengthen peacetime nuclear weapons security to the utmost. The available means range from Permissive Action Links (locking devices) and tamper-proof disablement to component removal, weapon destruction or evacuation, and improved guard facilities and site protection. The program is receiving high, well-deserved priority.

Summary

The foregoing examination of security arrangements in the NATO area, and of a number of key implications for U.S. forces and programs, illustrates the manner in which the role and structure of U.S. military forces may be shaped by such undertakings. A similar analysis for other regions (the South Korea-Japan area of Northeast Asia and the Western Pacific, in particular) developed on the basis of the underlying political and security linkages of these countries with the United States would round out this second major category of United States military endeavors.

We turn now to the third and final major category—U.S. military provision for its security in the "world beyond."

8

The World beyond—What Forces for U.S. Security?

In the world beyond, the seas have a special importance to U.S. security. In peace, crisis, and war, the dependence of the United States and its allies on free use of the world's seas gives rise to a wide spectrum of military requirements. In the shaping of the role and posture of our naval forces, these requirements have been undergoing profound evolution as the Soviet naval expansion has proceeded on an impressive scale through three postwar decades. Also part of this evolution are new instances of potential crisis and conflict that have developed and the invention of new technologies of missilery, nuclear weapons, and nuclear propulsion that have modified the available arsenal of naval ships, planes, and weapons.

American Security at Sea

We look to the naval services (Navy and Marines) to provide the access we require to the seas and to thwart or prevent any attempt by a hostile power to deny us the benefits of their use. The ultimate aim is to make full use of the seas for peaceful commerce, to provide a flexible means of responding to crisis with naval deployments and other measures of military support, and to deter and defend against forceful interference with our use of the seas.

Three principal contingencies give focus to our security concerns in this regard:

1. major hostilities with the Soviet Union that include operations at sea
2. localized interference or threatened interference with our free use of the sea for shipping or routine naval activity
3. crisis or conflict abroad in which U.S. interests, commitments, or responsibilities to allies and friends are involved

As in all cases of possible conflict with the Soviet Union, the specific American security interest we are here considering—safeguarding the free and beneficial use of the sea—will be best served by deterring and avoiding all kinds of U.S.-Soviet armed conflict (while preserving national values and valued activities) and by reducing the risk of clash in every practicable way through negotiations. Use of force, as in strategic nuclear hostilities or a NATO conflict, would be a last resort to avoid even worse consequences. However, it would

131

constitute failure in large degree to achieve the true aim of peaceful use of the sea.

Likewise, deterrence of localized interference, or better yet, peaceful resolution of the issues that might cause it, is certainly to be preferred to armed clash and the losses—to shipping and to naval forces—that would surely result. The inherent vulnerability of such shipping, fishing vessels, drilling, and other mineral exploitation—not to mention noncombatant naval activity—is such that secure and orderly use of the sea must be the norm; any attempt to provide constant and complete security against a troublemaker who can choose time, place, mode, and scale of attack would be utterly impractical. We need and want no more Pueblo or Mayaguez incidents, nor unilateral acts such as the Libyan mining of nonterritorial waters in the early 1970s.

Even in crisis situations such as the Arab-Israeli war of October 1973, U.S. interests are likely to be best served by arrangements or tacit understandings that prevent, if possible, our becoming embroiled in actual maritime hostilities. Provision of support and assistance without escalating or spreading the conflict, particularly to a point where clashes of U.S. and Soviet ships or aircraft could occur, should certainly be the aim. Freedom for Soviet ships to deliver arms and equipment may be the price we must pay for our freedom to do the same without triggering armed conflict between the two of us or between U.S. or Soviet forces and the Arab-Israeli forces locked in combat.

Nevertheless, while the practices of deterrence and detente are to be preferred, defense—actual naval warfare—may be required. And as in other arenas of action, a well-designed, ready capability for such defense is a principal essential underpinning for deterrence of resort to force and for negotiations under the label of detente to reduce or remove causes of potential outbreak of hostilities. Our chief concern is with the Soviet Union, the thought being that U.S. naval forces adequate to deal with Soviet naval might will be adequate and suitable, with minor adjustments or augmentations in ships, facilities, and supporting establishments, to deal with localized interference and requirements for support of lesser crises in which the United States becomes involved.

Opposing Force Capabilities

It has generally been considered that there is a significant difference in U.S. and Soviet dependence on the sea, particularly in case of war, and hence in the overall pattern of naval capabilities that each needs to maintain. While in peace both nations use and need the seas for commerce and for naval presence in support of international contact and influence, the Soviet Union and its major allies in wartime would not be dependent on sea lines of communications, as would be the nations of the West. Consequently, Soviet security is often thought to require only the denial of the seas to the West in case of war, while the

Western states must be able to achieve and assure large-scale, multifaceted utilization of the seas—a far more demanding requirement.

The respective naval forces of the United States and the USSR generally reflect these differing requirements, although the Soviet building programs have been extended, in the construction of the Kiev-class carriers and underway replenishment ships, to provide what may prove to be the beginnings of a more ambitious undertaking. Ultimately, it could conceivably mean a possible full-fledged worldwide blue-water wartime role. There seems to be no inherent reason—in Soviet industrial capacity or political purpose—to stop short of ultimate naval parity of capabilities. Until these new types of ships began to appear, however, the Soviet Union, in terms of the principal naval categories—carriers, surface combatants, submarines, and amphibious and support ships—had apparently contented itself with concentrating largely on the denial role in relation to its wartime naval capabilities. In major exercises and in the heavy emphasis given to surface combatants and attack submarines in their construction programs, they put priority emphasis on countering or challenging the U.S. carrier and amphibious forces that would project naval tactical air power and amphibious forces ashore. The same Soviet forces—surface combatants and submarines—give them at the same time a powerful capability against allied sea lines of communication and the naval forces that would guard them. In addition, these forces, together with the limited naval infantry forces of the Soviet Union and its allies, give them the ability to protect and extend their flanks, in the far north, the Baltic, the Black Sea and Mediterranean, and in the Far East. In the other traditional categories of naval action—long-range projection of naval power, establishment and protection of their own sea lines of communication, and distant and extensive antisubmarine operations—the Soviets are still much less advanced. Their ballistic missile submarines (previously discussed) constitute—like those of the United States—a further naval category of advanced technological capability and vast destructive power, separate and distinct from these general purpose naval forces.

In peacetime, and in crisis situations such as Angola in early 1976, these ships have provided the Soviet Union an ability to maintain a presence in far-flung areas of the world, to support client states such as Somalia or South Yemen, to protect the large Soviet commercial and fishing fleets, to maintain a widespread naval intelligence watch, and to counter or offset deployed American forces, in the Mediterranean for example.

For the United States, the carrier remains the core of U.S. naval strength—for projection of naval power, for sea control, and on many occasions for response to crisis. The maritime environment in which it operates is undergoing major change, however. No longer, in the presence of modern, lethal Soviet surface combatants and submarines, can the carriers be expected to operate from a privileged sanctuary, as they did in Korea and Vietnam, and as they would have done in the Atlantic and Mediterranean until recent years. Nevertheless,

American carriers and amphibious forces continue to provide a powerful capability for projection of naval power as a first role and mission—one still of great importance to security and stability.

The establishment and protection of sea lines of communication form a second charge on American naval strength. Fundamental to the NATO defense capability are early large-scale reinforcement and resupply, for which secure sea movement is indispensable. The naval forces required for sea control are taking on a higher priority, confronting the U.S. Navy with severe problems of force and ship design in a time when costs (and individual ship capabilities) have sharply risen, but large numbers of ships are still indispensable for this role.

We may note that this sea control task (to which the carrier forces can provide a timely and large contribution) would, where and as carried out, have a further value. It would deny the Soviet naval forces the ability to sustain a wartime presence or maintain sea lines of communication to other countries. American forces engaged in this task would counter and destroy the Soviet surface combatants and submarines assigned to attack our carriers and amphibious forces, as well as our sea lines of communication. Although time would be required for the purpose, particularly for progressively eliminating the Soviet submarines, such U.S. operations would aim at affording sea-borne tactical and logistical support for the collective allied effort.

From a sea-going force one-fortieth or less the size of U.S. naval forces at the close of World War II, Soviet naval forces have increased to a level that since 1973 has exceeded that of the United States in major surface combat ships (in 1976 approximately 220 to approximately 180). Thirteen of the American ships are attack carriers, however. These are unmatched in the Soviet fleet. In attack and cruise missile submarines, Soviet naval force strength is over three times as great as that of the United States (approximately 250 to 75); of these, some 75 of the Soviet submarines are identified as nuclear-propelled, against approximately 60 of the American.

The Soviet surface fleet includes large numbers of highly capable ships—well armed with offensive and defensive missiles of proven effectiveness, equipped with excellent communications and electronics packages. Their building program is emphasizing heavily armed guided missile cruisers, and modernization and replacement of older destroyers and escorts to provide a formidable surface capability. Antisubmarine capability is being significantly improved with the addition of the aircraft carriers of the Moskva and now the Kiev classes. Soviet naval aviation, capable of operating in close conjunction with surface forces, includes strike aircraft, reconnnaissance, and support aircraft. The importance attached to this role is demonstrated by the allocation of Backfire supersonic bombers, which greatly increase the range, speed of attack and strike capability of the naval aviation forces. The Kiev class aircraft carriers, equipped with fighter-type aircraft, will add capabilities for air defense, strike, close air support, reconnaissance, or ASW from more forward positions.

Implications for U.S. Forces

The increase and expansion of Soviet naval capabilities puts added stress on U.S. naval roles and on the ships and planes to fulfill them.

The "denial" capabilities of Soviet naval forces—surface combatants, submarines, and naval aviation—put them in position to contest U.S. projection of power and U.S. protection of sea lines of communication needed for support of NATO or of Japan and Korea in case of major hostilities. Beyond this, they can challenge U.S. "sea control" capabilities, which are a prerequisite to success in power projection and sea lines protection. Both quantatively and qualitatively, a U.S. response will be necessary. While the details are to be worked out by naval experts, the main lines of added effort will have to include more ships, especially escorts and probably additional attack submarines as well, strengthening of AAW and ASW through continuing incorporation of the most advanced technology that can be devised, and emphasis on ships and weapons—especially guided-missiles—that can fight on the surface against the best the Soviets have to offer. Many programs in recent years have included substantial efforts in all these fields. The Harpoon missile has been long awaited and sorely needed. No let-up in need can be foreseen; to the contrary, continued high priority will be required. Rational resolution of the high-low mix in every category—surface combatant, submarine, and naval air—is an inescapable need, given the requirement for substantial numbers of ships and aircraft, and the necessity for some of these—but in most cases, only a part—to be of the highest technology and capability attainable.

If the Soviets should pursue their naval expansion to the point of seeking to establish the whole range of naval capabilities—power projection, sea control, and sea lines of communication—a whole new strategic era would open up for the United States and its allies, in terms of their dependence on the sea. Already such Soviet sea control capabilities are emerging and expanding in limited areas particularly vital to their security—the Barents Sea and White Sea, the Baltic Sea, Black Sea, Sea of Okhotsk and Sea of Japan. Their capabilities around and beyond the North Cape of Norway—out to the Greenland-Iceland-United Kingdom gap—are increasing, as is the threat they pose to the Eastern Mediterranean basin. Should they attain assured base facilities on the Adriatic, or gain a position on the Aegean in war, they would be in position to contest for sea control in the Eastern Mediterranean. With North African bases they could contest for sea control in the Western Mediterranean as well.

None of this reduces the need for U.S. and allied naval strength and modernization. Rather it reinforces the requirement for quantitative and qualitative improvement. In addition, major Soviet progress in this direction would, at the least, require an examination of alternatives already in use, including the prestocking of heavy military equipment for additional units, increase in stock levels of ammunition and other supplies, additional land-based tactical air, expansion of airlift carrying and unloading capacity, and similar measures to ease early vulnerabilities. While these measures tend to be more costly and less desirable than presently planned heavy reliance on maritime operations, they will be the inevitable consequences of such a heightened Soviet maritime capability on the high seas.

American Security and the Lands beyond the Sea

In addition to the need for free use of the sea stands the possible need to act in the lands beyond. However, the lessons of recent years—in Vietnam, in particular—tell us that the list of countries has been greatly shortened in which U.S. military commitment is a serious possibility in the present state of world affairs.

In recent years we have seen in nations as far separated as Nigeria, Argentina, Lebanon, and Bangladesh that ours is a world where violence is still endemic. Seizure of power or insurgency within countries, attempts at secession and partition, foreign penetration and support of warring factions—all these highlight the present scene, and seem certain to persist. Ambition, xenophobia, historic hatreds and other emotions, territorial and other grievances, tribal consciousness, and ethnic and religious divisions weigh at least as heavily as the urge for peaceful progress, economic advancement, public safety, and freedom from the devastation of war.

These then are areas that seem destined to be the locus of turbulence and instability for years to come. As well, they are areas in which U.S. interests and Soviet interests—particularly Soviet interests in installing communism and Soviet hegemony—are likely to prove conflicting. But they are at the same time areas in which U.S. military intervention is likely to hold little attraction for the American government or people. It is to other instruments of international influence and intercourse—and in most cases, *only* to such instruments—that our leaders are likely to turn. Nevertheless, there do exist more than a handful of nations that are important enough to U.S. security and well-being—or have strong enough ties with the United States of other kinds—to pose the possibility, even if slight and uncertain, of potential military involvement.

A broad category of nations possessing this special significance can be identified without too much difficulty, even though one might debate the precise listing. The Middle East continues as one such area. The fate of Iran, Saudi Arabia, Kuwait, and the Arab Emirates, with their present production and future reserves of· oil, cannot be left wholly to chance. Indonesia, Malaysia/ Singapore, and Australia and New Zealand must also be recognized. Special bonds and a variety of commitments would add Israel, Spain, Morocco, Nigeria, Taiwan, and the Philippines to the list. Our hemispheric partners—especially Brazil, Mexico, Venezuela, and Chile—have a special significance.

At the other end of the scale, however, little can be seen in current American attitudes that would suggest, for scores of nations in major areas of Asia or Africa, any serious possibility of American military involvement. A Soviet communist campaign of takeovers by force that began to make major inroads, organizing these areas into a new empire hostile to the West, could stir a deeper concern and new decisions in the West. On the evidence, the Soviet initiative in Angola did not raise such a concern.

The present attitudes probably reflect a number of elements. There has been a marked decline in the perceived importance of such areas to the United States. Also, they are seen to be less in need of U.S. support than in the early postwar years when they were emerging from war's disruption, or from colonial rule, and when the United States was seen as predominantly powerful, both economically and militarily. The rejection by so many of Western democracy in favor of authoritarian rule has been discouraging to Americans who felt an obligation to try to offer the opportunity of widely based self-government.

With the exception of Israel, and possibly one or two others such as the Philippines, it seems likely that it would take the prospect of drastic and imminent harm of a material nature—interference with the flow of oil, or possibly copper or bauxite, to be specific—to occasion any serious thoughts of U.S. Military Intervention. A Pueblo or Mayaguez incident cannot, however, be wholly excluded, nor flagrant terrorism, as other possible triggers of military action.

Accompanying and reinforcing the shift of attitude just mentioned is a decline in the perceived efficacy and utility of limited war. Our capability to wage such warfare or to conduct lesser operations of a counterinsurgency character is assessed much lower than in the heady, pre-Vietnam days. In particular, the difficulty of bringing such involvements to an end, and disengaging from such commitments, has the evidence of Korea and Vietnam to give it emphasis. Interestingly, the resulting inhibitions that now work on the United States do not work so strongly on the nations that are directly affected, as the Arab-Israeli, Indo-Pakistan, Bangladesh, and other conflicts attest. In minor part, this reflects a sense (not wholly true) that we lack the kinds of arms and tactics that would be applicable to limited conflict. In greater degree, however, it reflects inability at the level of national leadership to form policies and to gain and hold national support for military plans and operations that would achieve success in such limited conflict.

In any case, the lack of capability at the middle range of military conflict and the strong opposition to any such involvement in foreign lands have put a premium on nonmilitary methods of dealing with such issues. In a sense, the military power we have is questioned as being almost unusable for these purposes. Of far more immediate relevance and benefit are the processes of political influence and diplomatic assistance, for example, in easing or removing conflicts and problems with neighbors or antagonists; of economic cooperation through aid or improvement in the terms or the flow of trade; of technical aid, especially in modernizing and developing the economy; and of cultural/psychological efforts to build a sense of friendship and a basis of reciprocal respect for each other's achievements. Only if such means should fail, if clear U.S. interests were deemed to justify it, and if operations that promised success within costs our people were likely to accept could be devised, would military intervention be likely to receive serious consideration.

The foregoing discussion has noted two broad and important classes of possible contingencies that might conceivably evoke U.S. military involvement (and which provide some basis upon which to develop and maintain U.S. military capabilities). The first would be hostile action resulting in a halt to the flow of oil or other raw materials vital or nearly so to the American economy and society. A factor that might increase the possibility of American action would be clear evidence that such action involved military forces from outside the country concerned, or forces from a small insurgent faction within, strongly supported from outside. The second contingency would be an accumulating sequence of takeovers by force on the part of Soviet-based communism sufficient to establish a pattern that, if continued, could choke and render powerless the Western industrial democracies.

It is evident that neither of these contingencies holds high place in current Western thinking. Nevertheless, taking a longer view, it must be noted that both contingencies are clearly consistent with possible scenarios by which the Soviets might seek to carry forward the processes of world revolution to which they are so explicitly committed.

These are means by which "national liberation movements," supported by the forces of socialism in the USSR and its principal allies, could intensify and exploit what they call the current "crisis of capitalism." The major questions for them would be those of timing and technique, i.e., whether the "correlation of forces," which they declare to be shifting in their favor, had reached a point that made the possible risks and costs of such action acceptable to them.

At an earlier point in this study we attempted to identify the principal interests and the policy objectives—under the headings of defense, deterrence, and detente—that are most closely associated with American security needs and possibilities in the third world. In summary form, they included:

support (within practical limits) of national freedom and of strengthened indigenous nationalism within the third-world countries

safeguarding the freedom of the seas

striving to maintain at least an essential minimum of order and stability requisite to expanding trade, commerce, economic development, and well-being, in an era of increasing economic interdependence denial of Soviet territorial aggrandizement, of seizure and control of free-world resources, and of extension of their influence and control into foreign lands that would carry with it serious hazard to the United States. They must in particular be denied control over Middle East oil vital to Western economies (including that of the United States).

In fact, the interest of the United States in the availability of Middle East oil has now become so great as to make intolerable any major interruption of its flow, particularly if imposed by force with hostile intent. Oil has without question assumed a unique status and importance in this respect. The full consequences of the oil situation have yet to be recognized or experienced.

As repeatedly noted, there can be no certainty, particularly in the post-Vietnam, post-Watergate period, as to whether the American government will or will not act militarily in support of these or comparable interests, should they be violated. But to retain the option of choice, recognizing that the decision in all likelihood would only be made in the light of the specific circumstances of a specific crisis, threat, or challenge, military capabilities of the nature that would be needed if a "go" decision were taken would have to be available. It seems reasonable then to assume that our government and people will support the retention of such an option. This is true particularly since many—although not all—of the military means required are made necessary by the needs of regional security (in NATO and the Japan-Korea area) and for security at sea. Accordingly, we can proceed to examine the implications and requirements of defense, deterrence, and detente policies in this regard.

For defense, the requirements are:

providing a capability for intervention operations (typically of limited scope), possibly coupled with counterintervention against Soviet or Communist Chinese initiatives, and/or those of their proxies, such as the Cubans.

maintenance of sea control (as previously discussed).

For deterrence, the requirements are:

where alien initiatives with possible hostile intent might approach or do heavy harm to areas vital to the United States (including, specifically, the American continent, Western Europe, Japan/Korea and the Middle East), demonstration of U.S. will and ability to oppose such action by military means, if necessary

making clear U.S. readiness to safeguard and restore the freedom of the seas, if threatened or violated

For detente, the requirements are:

140

giving our positive support to the development and implementation of an international Law of the Sea that underpins the freedom of the seas

supporting, and extending where beneficial, the arrangements on safety at sea, including avoidance of provocative naval tactics

strengthening commitments for consultation in case of local disputes or clashes involving or impinging on the United States—such consultation to be conducted with the Soviet Union and Communist China, where they are involved, and with the other particular countries concerned, whether the Soviets and Chinese are involved or not. The object would be to localize the dispute, and prevent its uncontrolled escalation.

building up a "structure of peace"—a network of understandings and reciprocal undertakings beneficial to both sides, for their "linkage" or "carry-over" effects—such as they may be—tending to reinforce stable and constructive relationships with the United States.

If military operations should be undertaken in support of these interests and policies in the future there will be a need for more effective methods of higher direction than have yet been developed or demonstrated. This is particularly true if the involvement is to be very long or very large. Of recent experiences, the U.S. intervention in the Dominican Republic in 1965—although challenged as to its necessity and justification—was carried out without excessive destruction or loss of life. The military operations confined the violence and set the stage for political efforts to find a solution. The experience in Vietnam, setting aside the fatal collapse of morale and support in the United States, indicates some of the key issues and tasks that an intervention of substantial size, duration, and complexity is likely to encounter, and some of what may be needed to deal with it effectively at the operational level.

The Vietnam example should suggest to us—if proof were needed—that we are far from well prepared at the national level to understand, shape, and direct a limited military intervention of the kind that might be required. Accordingly, a review of some, at least, of the key issues that might be encountered may be of value. Among the issues to be dealt with (the list is by no means exhaustive) the following are likely to be included, if the Vietnam experience is illustrative of problems that may arise:

1. A clear working "definition of success" is needed, i.e., a postulated outcome that constitutes an acceptable minimum threshold in relation to our perceived national interests—one that people and government may be expected to support. Part of the confusion over Vietnam came from lack of clarity on this point. In countries of the third world—as in Vietnam—it will often be the case that continuing border hostilities and some measure of internal resistance must be accepted to continue indefinitely. Despite such residual struggle, the main life of the country—economic, political, and social—can go on in ways that the

people can and will bear in support of their national freedom and sovereignty. If a full cease-fire in a limited time period is the only end acceptable from an American standpoint, then the attacking party—like the North Vietnamese in that case—would have only to prolong and protract the conflict in order to wind up the victor, unless the United States was willing to attempt to force a full surrender, through invasion of the seat of the attacking power, for example. If a limited form of success is not acceptable, then the decision to become involved becomes a gamble that the other side will weary of the struggle before we do, and the whole enterprise becomes one of dubious wisdom. For operations of this kind, there is special value in thinking through from the outset what definite limits may be appropriate. Likewise, it is necessary to face up to "worst cases"—the situation in which we will find ourselves if initial efforts thought likely to succeed do not in fact succeed.

2. Subordination of collateral purposes to the aim being sought by military action becomes necessary. If young Americans are to be exposed to the hazards of combat, the cause should be kept to the highest-priority of national need, and methods of combat should not be imposed that will add to casualties or delay the attainment of conditions approximating the "definition of success." Clearcut channels of responsibility and authority become an important obligation of the U.S. president and the national command authorities (i.e., the Secretaries of State and Defense, and the Joint Chiefs of Staff).

3. A considerable number and diversity of efforts are likely to be involved, and a considerable coordination and orchestration of these efforts may be needed. In Vietnam, for example, action had to be pursued concurrently on many fronts—against high-capability North Vietnamese divisions, lower capability local and guerrilla forces, lines of communication personnel, and "infrastructure," i.e., the hostile politicomilitary organization. A whole panoply of allied forces, ranging from our main-force units down to local regional and popular force garrisons, had to be employed against the diverse enemy components wherever they were located, in terrain ranging from heavily populated and cultivated lowland, coastal, and delta regions to more open piedmont, high plateaus, mountain jungle, and coastal or other swamps, at varying distances and with varying access to the sanctuaries and support areas in Cambodia, Laos, and North Vietnam.

4. There may be need to pursue operations to a point that will destroy an enemy's ability to fight, i.e., to go far beyond efforts simply to induce him to stop fighting by targeting his will to fight. If the final North Vietnamese victory taught anything, it is the power of persistence despite vast losses in the communist-controlled nation.

5. Special effort will be needed to acquire a valid understanding of the area, the people, and their culture. In Asia, for example, there are difficulties deriving from the profound differences in cultural and psychological heritage. As Americans, our learning years in school were strongly shaped by European and

Atlantic modes of thought. Among the differences that come into play and have to be respected àre differences in the things the people value and in the attitudes they hold toward their goods, their land, their houses, their families, their religions, and their associations with friends and fellow-townspeople. There are differences in person-to-person transactions and relationships, in attitudes toward authority, in attitude toward human life, and in the sense of nationhood and self-rule. Nor is the base of military information so familiar to us—such matters as distances, strengths of forces, levels of skills, logistical capabilities, terrain difficulties, or the stamina and morale of their troops. The trial of arms in Vietnam has left us with evidence as to the relative fighting strengths and weaknesses of Asian military forces—among themselves and in relation to our own. Even so, however, major uncertainties and unknowns will abound, in any intervention operations that may be undertaken, by virtue of differences in weaponry, in organization and in methods of combat.

6. The task of estimating requirements—of troops, supplies, money and time, in particular—is beset by special difficulties of limited experience, conditions difficult to assess and allies hard to evaluate. Yet such appraisals are indispensable to informed judgments and decisions where U.S. military involvement is under consideration. The enemy forces that may be encountered, and the scale and continuity of logistic support they might receive—ammunition, POL, and replacement equipment—will be of key importance to a determination of the scale of effort that might be required of us.

Much in the foregoing analysis tends strongly to support the long-standing but twice-violated injunction not to get involved in a land war in Asia. The hazards of assessment and prediction are especially great, the friendly and hostile environment inevitably highly uncertain. Yet Vietnam suggests, it may be argued, that if sufficient reason for such intervention exists and is so evaluated by the American people, and if they could be provided a reasonably accurate picture, in reasonably balanced perspective, of the actual situation and its development, a successful outcome would be possible. Now that we have burned ourselves so badly in Vietnam, the vast difficulties, and the uncertainty as to whether such a valid picture can or would be provided, should convince us that such enterprises—whether in Asia, the Middle East, Africa, or even Latin America—are likely to tend to be reserved for crises and challenges posing only the most crucial threats to the United States for a considerable time to come. In the world of the late 1970s only something like an attempt to use an oil embargo as a weapon of conscious and deliberate economic warfare against the West—attacking millions of Americans at the gas pumps—would appear likely to prove sufficient cause to swing the United States away from "No More Vietnams." A threat to the existence of Israel might evoke the same reaction; no one can be sure. In any case, a strong emotional element widely shared in the United States would seem to be a requisite to any response that went beyond such things as action at sea, initiatives not involving use of U.S. military forces in combat, and direct approaches to the Soviet Union and Communist China, if they were significantly involved.

Implications for U.S. Forces

We do know, though, that sudden, sometimes unpredictable challenges to American security and well-being may arise, that the American people have immense reserves of strength to deal with them, and that effective military response will be demanded if any fighting begins. For these reasons—whatever the mood of the country may have appeared to be in the immediate post-Vietnam era—and to provide a deterrent to the kind of foreign adventures and challenges that might trigger such U.S. response, it would be imprudent not to maintain the kinds of forces that could provide the kinds of intervention capabilities that might reasonably be demanded. Land forces—Marine and Army—and tactical air forces, both sea-based and land-based, supported by strategic sealift and airlift, provide military force of the kind needed. Capabilities for sea control to give operational freedom and afford usable access to the crisis area, and to permit projection of power ashore (discussed at an earlier point in this chapter) round out the main military requirements.

For action in the regions of the third world, major uncertainties nevertheless remain as to where and when our forces might be needed, and just what they may be called on to do. Force design must deal suitably with such unknowns, and it must provide military power that is capable of responding to a wide range of requirements in diverse environments against diverse kinds and scales of resistance. The task, in professional terms, is difficult; it is by no means unmanageable. In fact, for reasonably limited costs and burdens, force packages of combined arms of a size and composition giving good insurance against a wide spectrum of contingencies can be built and maintained. As indicated earlier, it is the capacity at the national level to provide top-level direction that is more likely to prove a source of difficulty.

The uncertainties regarding possible third-world contingencies put a premium on flexibility and promptness of response in the forces that would be used to deal with them. Quick-reaction Marine and Army airborne units have these qualities to a high degree. Carrier air and land-based air (where bases within range are available) can provide concentrated air support that might be needed to gain immediate overall superiority. Prompt followup forces may be needed. They can come from forces having a general reserve capability and role—the Marine and Army divisions and the tactical air forces in the United States, in particular. These are units of the active forces, mostly maintained at high levels of readiness. For their delivery to the target area, airlift will need to be used immediately with sealift following as soon as possible, but necessarily at a considerably later date. Should force requirements during an intervention continue to expand, Army, Marine, and Air Force units from the reserves can be and must be called up—whether for use in the crisis area or to restore the general reserve capability for support of NATO or the Japan-Korea area. Obviously, high-level decisions may be required at many points as to the relative priority to be given to the urgent needs of the crisis effort as against a need for possible NATO reinforcement. Although this is perhaps less immediate, it may nevertheless be more telling in its potential consequences.

A similar premium, as indicated, is placed on strategic airlift and sealift.

This is not the place to attempt to prescribe in specific detail how the general reserve role for active Marine, Army, and Air Force units—in the continental United States, Hawaii, and Okinawa—can best be accomplished. A few key desiderata may be suggested, however.

The quick reaction forces should be just that. They may be of limited size—a reinforced battalion, for example. They should be capable of entry into target areas, by force if necessary, in as little as twenty-four to seventy-two hours. That means that carrier air, or land-based air if suitably positioned, should be able to conduct preliminary or preparatory attacks within the same time frame and to continue to do so for as long as required after the forces on the ground had been introduced. Communications, command, and control, as well as intelligence and logistic support, would have to be established concurrently.

The followon forces present a more complex problem. They must be suitable for employment in contingency areas where lightened scales of weaponry may sometimes be more suitable than the whole panoply of armor and artillery. But they must also be considered in the context of other commitment and reinforcement roles, since the United States is unlikely to maintain specialized forces of division or greater size that are not fully capable of serving in the NATO (or Japan-Korea) environment. This means they must basically have the capability for medium to heavy combat, as well as a high order of tactical mobility. Moreover, they must be prepared to operate within large, integrated forces.

The ability of Marine divisions to operate in relatively heavy environments— in NATO, for example—may require that they be augmented by heavy firepower, and mechanized mobility for the riflemen. Even on the flanks of NATO where commitment of Marines, perhaps in an amphibious role, may be required—in Norway, Denmark, or Greece or Turkish Thrace, for example— armor, artillery, and mechanized mobility would be sorely needed. The Army's Airborne and Air Mobility divisions would require much of the same. In addition, some of the Army's heavier divisions should be capable of operating with reduced scales of heavy combat equipment—artillery and armor in particular. However, there is a trap to avoid here: namely, stereotypes such as "Vietnam is not a place for tanks." Specific terrain analysis is required, and wherever modern weapons technology can be employed instead of reliance on human resources, the lives of our military can be saved, and the conflict can be brought to a quicker end.

The Forces Available

U.S. Army divisions in 1976 numbered sixteen active and eight reserve. Eleven of the active divisions were in the continental United States and Hawaii, seven of them (and five of the eight reserve divisions) of light varieties, i.e., infantry, airborne, or air assault. Pentagon plans called for two of them to be brought to heavy status (i.e., as armored or mechanized divisions) within the following few years. The Marines maintain three active and one reserve divisions, with one of the active units on the U.S. East coast, one on the West coast, and one in Hawaii and Okinawa. Each Marine division is coupled with a powerful tactical air wing to provide a composite land-air "Marine Amphibious Force." Amphibious lift provides total capacity sufficient to transport just over one division/air wing; this lift has normally been deployed with about one-half in the Atlantic and one-half in the Pacific. Approximately two Marine battalion-size units ("Marine Amphibious Units") can be kept afloat continuously. Normally one has been kept (complete with helicopters) in the Mediterranean and the other in the Western Pacific.

The Air Force, Navy, and Marines, in combination, provided a total of some four thousand fighter and attack aircraft in 1975. Although totals are declining as some of the newer, more capable models are introduced, no major reductions are planned, and the advanced capabilities of the new aircraft give the total force a considerably enhanced combat power. A substantial fraction of the total is deployed to Europe, Korea, and Japan, but well over half of them remain undeployed and available for flexible commitment. While the Marine air units are primarily dedicated to use with their associated divisions, they are capable—if not required to be used in that way—of taking part in integrated tactical air operations and organizations. Although Marine authorities have been resistant to such use in the past—in Vietnam, for example—their air units demonstrated there that they can perform effectively in such a role. Considering the need for Marine forces to play a full NATO role where and as required, this is a capability that should be further practiced and developed.

The backbone of military strategic airlift is an inventory of some 300 C-5 and C-141 aircraft, the former numbering about 70. They are backed by nearly 250 long-range aircraft of the U.S. commercial airlines—the so-called Civil Reserve Air Fleet (CRAF)—some 90 of which are passenger-only aircraft, the remainder being capable of carrying cargo. This arrangement has proved an excellent response to airlift needs. It can be substantially enhanced, at limited cost, by proposed programs to "stretch" the C-141 to enable it to carry cargo of greater bulk, to provide the C-141 with an aerial refuelling capability (increasingly important as use of foreign bases becomes subject to doubt), and to modify the CRAF aircraft to handle military cargo more effectively. By such important qualitative improvements, U.S. capabilities in the crucial initial days of crises can be usefully enlarged.

Sealift would be essential for movement of military equipment and supplies in large quantities over a sustained period. As noted, U.S. amphibious shipping could lift the assault elements of just over one Marine division/air wing team, with their supporting elements. Beyond this, major reliance would have to be placed on sealift in case of a military undertaking of substantial size and duration. Defense sealift in active use in peacetime is severely limited in capacity. Additional sealift from commercial shipping lines or from a modest-sized fleet of mothballed ships capable of quick reactivation would be needed. A good deal of further development of programs along these lines will be necessary if an appropriate prompt sealift capability is to be made a reality on a standby basis.

In general, the basic military assets required to provide what might be needed for third-world undertakings appear to be amply available. Qualitative improvements, heightened readiness, and enhanced ability to assure prompt and sustained access to the critical area, to provide command and control, as well as communications, intelligence, and logistical support—these are the more important needs, together with the ability to tailor the force quickly to the particular combat or crisis environment. On the other side of the coin, forces particularly suited to operations of this kind should also be capable of reinforcement and reorganization. Thus they would be able to perform in a higher intensity environment such as NATO Europe, where heavier firepower and higher tactical mobility are required.

By far the most important requirement exists, however, at the level of national decision and direction of such an effort. At a time dominated by the rule of "No More Vietnams," such needs tend to be masked and neglected. If intervention should be seriously contemplated or undertaken however—whether in the Middle East, Latin America, or elsewhere—such needs would immediately pose one of the most demanding sets of tasks to be faced. Faulty handling could jeopardize the success of the endeavor, as we saw in Vietnam.

9 A Concluding Comment

As we come to the end of our analysis and assessment, it is appropriate to ask what results such an effort can hope for in the real world in which we live. It was noted in the Preface that there are many who doubt that national values can be made to serve in this way as a basis for policy and force determination. Similar doubts attach, in fact, to every succeeding step in the analysis. Two overall questions emerge in particular:

1. What are the chances that an approach of this kind (proceeding from interests, to policies, to force structure and major weapons systems), can actually be put into practical use? If it can be, what is the likelihood that evaluations and conclusions along the lines of those put forth in the preceding chapters might actually be deemed sufficiently valid by responsible decision-makers to serve as a principal basis for military budgets and action programs?

2. How likely is it that international events and circumstances will in fact develop along the lines described or foreseen in this discussion? What major changes in the world security environment, even if not deemed likely, are nevertheless sufficiently possible and so potentially harmful (if they were to occur) as to warrant our giving them our attention, even if only as remote contingencies? In this connection, are there changes we should expect in the role and priority of our traditional security concerns, and in the foreign problems that confront us? Are there likely to be changes in the relationships of major world nations that will upset and endanger the world security balance?

As to the first question, there exist major doubts and resistances, both within government and without, about being guided by such a pattern of analysis as has been presented in this book.

The arguments and impediments—voiced and unvoiced—to the process of determining force structures by the method just outlined are numerous, and in many cases substantial. It may be argued, with some reason, that the matters dealt with are too complex to be subjected to such a process of analysis—that there are too many unknown and unknowables. It can be argued also: that the process proposed is itself too complex and that a simpler basis, such as a straight matching of Russian strength, would be more feasible and appropriate; that determinations thus made will be too inflexible to meet actual needs, or if left flexible will not be definitive enough to shape programs; that such an approach is too deterministic, allowing too little scope for the play of human interests and preferences; that techniques for converting policies into programs are inadequate. Arguments have been voiced that: purpose-driven solutions are too much

147

a blank-check, disregardful of costs, that in fact will never be written; foreign policy is too volatile, multifaceted and unpredictable to serve as basis for programs of force building that require continuity in such matters as personal recruitment and training, equipment design and procurement, installation and communication development, etc.; the values served are too vague, shifting and incalculable to be used for such a purpose; and policies such as the Nixon Doctrine, or the so-called Guam Doctrine enunciated later, must be couched in such general terms as to prohibit the derivation of specific goals and programs from them. Finally, it is argued on practical grounds that so many divergent interests and interest groups are involved as to leave it impossible to obtain agreement—meaningful agreement, that is. The inability of the government to produce a "Basic National Security Policy" document in the Kennedy and Johnson years is cited as conclusive evidence for this point.

Together, these arguments—many of them overlapping, some of them contradictory—have a weight that obviously must be recognized. Yet they just as obviously cannot be allowed to suggest that our national security interests and policies should be disregarded in establishing military force structure. Rather, they should serve to introduce useful refinements, caveats, and limits to the application of this method. For this purpose it is appropriate to take up the arguments in turn.

The argument that the matters dealt with are highly complex must be granted. Yet the extreme alternative—to not mold policies and programs to national values and interests, but to rely on a beneficent "hidden hand" to bring favorable results from the kinds of interservice quarreling, logrolling with Congress and industry, and loosely controlled military posture to the outside world that we have seen too often in the past—is obviously not a serious possibility. The complexity and uncertainty of much of what the future will bring place a high premium on flexibility—on forces and policies that are applicable to a wide variety of security challenges and strains, in many areas of the world, together with some multiplicity or even redundancy of approaches and types of forces to assure that all eggs are not put in a single basket. The strategic nuclear Triad—bombers and land- and sea-based missiles—targeted according to a variety of options; the diversity of NATO forces—land, sea, and air—committed to a doctrine of flexible response; and flexible, ready forces for sea control or quick reaction to crisis all respond to these needs. The factor of uncertainty tells us that we have to look for *hedges* against the unknown. In military planning (as in diplomacy) this is not an unfamiliar task, since adversaries normally enjoy a considerable range of choice, which we regularly recognize and respect in the range of options and responses that we pursue.

The contrary argument—that this type of analysis is itself too complex, and that a simpler path must be used—has value only to a degree. The idea of a raw balance with the Soviets—a tooth for a tooth, a claw for a claw—will not suffice. It is true that Soviet military strength in the strategic nuclear forces does indeed provide the primary basis for determining our own. But in the NATO area, the West and the Warsaw group each has its own interests, different from each other in major ways, and the geography that each has to work with likewise is different; force needs are, therefore, also quite different. Elsewhere in the world the differences between American and Soviet aims, ambitions, and modes of action mean that direct comparisons of forces are of limited value.

The argument that the arrangements that result from these processes will necessarily be too inflexible can be partly dealt with by deliberately seeking to design flexibility into our policies and programs, as previously suggested. It is, of course, true that a certain degree of inflexibility results from any specific force structure that is adopted, whether it is designed from the top down or the bottom up. But budgetary pressures are in any case going to force choices to be made, and the argument made herein is that the dominant—although *not* exclusive—voice in making such choices should be that of the authority with the widest range of policy vision and decision responsibility.

The argument that this method is too deterministic similarly must be questioned. Popular interests and preferences in fact enter the determination at a multitude of points, most importantly and properly at the very beginning, in the identification and the articulation of the national values and interests on which policy and force determinations are based. And beyond this, Congressional views and actions and public opinion expressed by the press, by academic circles, and by articulate interest groups will certainly have full and continuing play at every point of the chain from commitments and policies to forces and weapons. The explicit analysis set forth in the preceding chapters has as one of its purposes to provide for such participants a target for examination and challenge. Moreover, determinations of this kind can never be rigid or static. The annual cycle of defense authorizations, fund appropriations, and budget reviews, together with the crisis focusing of public attention on the condition and the capabilities of our armed forces, assures that the process in actuality remains vital and dynamic.

The argument that techniques for converting policies into programs are inadequate is, of course, unchallengeable. But this is by no means a pure desert. Many powerful techniques *do* exist—in use consciously, or merely in practice— and the method that has been employed herein serves to focus attention on specific areas of policy development and application where gaps and short-comings exist and where improvements are necessary and by no means impossible.

The argument that no blank check will be written for security policies and military programs that are purely purpose-driven is likewise valid. It reflects the need for a kind of feedback that introduces a major sophistication (and inescapable complication) into the decision process for policy formulation, program setting, and budget preparation. The costs and risks for specific security policies and for specific military roles, strategies, and structures are, in practice, carefully estimated before the fact and repeatedly reviewed during and after the fact. They are then fed back into the process of defining interests and setting working goals, along with appraisals of the benefits anticipated or received—to revise and refine, through successive reformulations, the whole cycle of interests, policies, and military forces.

The argument that foreign policy is too changeable and diversified to serve, per se, as the basis for force determinations likewise has considerable merit. Although such foreign policy is in flux, it is not wholly so. There is much that is enduring and in fact responds to the national values and interests that provide the starting point for our analysis. The military forces, which show a slower tempo of change than the day-to-day foreign policy activities, can be (and should be) aligned to a kind of "envelope" of foreign policy situations and needs. In it the common and enduring elements of many likely or typical situations would be aggregated and defined—and then augmented with some safety factor for the expected margins of error or uncertainty that will inevitably be present in our intelligence calculations. This is precisely what we see in the support that our nuclear forces, and our NATO forces, have given to the main structure of American foreign policy since the end of World War II. In a way, this problem is the reverse side (i.e., our side) of the familiar military issue of intentions and capabilities—the former subject to swift change with changes in national policy, the latter much more stable and enduring.

The argument that national values are too vague and ephemeral, and comprehensive doctrines (such as the Nixon Doctrine of the early 1970s) too general highlights part of the real problem. The approach employed tackles the difficulty head-on, however, since it proceeds—as explicitly as possible—from values and interests to increasingly specific policies and even more specific force structures that express the values and interests in more concrete and reviewable terms—oftentimes quantitative. The resulting policies and programs bring out the risks and costs to which governmental decision—Executive and Congressional—and public opinion must be directed. And the practical difficulty of achieving the necessary common understanding and agreement is no more than a recognition of the deep importance and far-reaching ramifications of these issues.

There are, in sum, substantial and relevant problems in the approach that has been used. But they are inherent in any approach that attempts to relate action to purpose. The method that has been followed appears to offer the advantage of openness—an inestimable value in our democratic system at all times and an especially important value at a time when major changes are occurring in the pattern of our security problems and needs and public confidence in the processes and performance of government has been at a low ebb.

In this connection, the Congress itself has given a timely and useful lead, in pressing for the rational linkage of our defense force programs with foreign and security policy objectives. Section 812 of the Fiscal Year 1976 Department of Defense Authorization Act requires that the Secretary of Defense, after consultation with the Secretary of State, shall prepare and submit to the Committees on Armed Services of the Senate and the House of Representatives a written annual report on the foreign policy and military force structure of the United States for the next fiscal year, how such policy and force structure relate to each other, and the justification for each.

This was surely a useful step, throwing needed light on this traditionally—and inherently—weak link in government operations. For it must be acknowledged that the correlation of defense programs with foreign policy and national security policy has rarely been as close or explicit in the past as might be desired. The 1976 Report of the Secretary of Defense to the Congress responded to this requirement in an initial way. It may be expected that future submissions in this series of annual reports will develop and apply the technique more fully and explicitly.

The progress that is made in this direction seems sure to be of value in strengthening public understanding of the rationale and the need behind specific defense proposals and budgets, and it offers a useful possibility for adding needed public confidence in such matters. The effort should at the same time provide a useful tool for identifying redundant, outdated, and low-priority defense activities and burdens, as well as areas of military insufficiency or weakness.

One possible improvement is readily apparent. The value of the technique would be greatly increased by extending it beyond a single fiscal year, since many military programs take years to develop, and can only be dealt with intelligently in terms of a longer time-cycle—research and development activities are illustrative. This is a change that clearly needs to be made. However, there should be no illusion that the sheer task of building the logical bridges between policy and programs will be simple or easy. Military forces as different as ICBM forces, worldwide naval forces, and the land-air combinations in NATO Europe or Korea are vastly complex aggregations of military resources and skills, which provide combat capabilities that are sharply dissimilar in timing and mode of employment, area influenced, kind of protection provided, destructive potential, and costs and burdens incurred. What the diverse forces contribute to the support of policy varies greatly in the directness of the contribution made, and in the susceptibility of such contribution to exact measurement. Because lead times are long for the shaping and reshaping of military power, it seems certain that there will be a premium on foreign policies—security policies in particular—that have considerable endurance, or that can be supported by forces capable of flexible application. Nevertheless, the call of the Congress in its 1976 resolution has provided a useful initiative and a useful impetus for work that should, over time, have a constructive effect. Otherwise, program setting and budget making will surely, in the future as in the past, be in too many cases superficial, piecemeal, arbitrary, and self-serving. It is a part of our tradition that our people—and the institutions that serve them—once they understand and accept the job to be done, will mobilize the effort and make the individual commitments of interest and priority that are required. One can accordingly be optimistic that efforts along the lines of the present study, aimed at making the whole process more visible and explicit, will find a growing place in our governmental operations.

In that process, several of the points noted in the preceding pages in considering the arguments and impediments to the approach employed herein are important enough to provide working principles. They include, as major examples, the reach for flexibility; the inclusion of hedges against unknowns, miscalculations or sudden policy changes; and the provision for Congressional and public review at many points. In particular, it should be reemphasized that too rigorous a solution should not be attempted. If we were to require of ourselves unchallengeable proof that each element of military force will be required in specific operations at specific places and times, we would unduly bind ourselves. There can be no masterplan so detailed and precise. Rather, flexible forces capable of responding to a wide range of cogent potential needs should be the aim, whether we are speaking of strategic nuclear forces, regional forces in NATO, worldwide naval forces, or other forces for possible use in third-world contingencies. Broad balance can be as important as specifiable use.

It is of interest in this regard that as early as 1970 the doctrine set out by President Nixon in his "A New Strategy for Peace" provided one of the fullest frameworks for such determination that had yet been seen. It was supported organizationally by a special committee established within the National Security Council structure, the "Defense Programs Review Committee." Yet, except for a very few significant actions in the initial period after its creation, this committee—ideally situated to correlate power and purpose, under a president closely concerned with international questions of just that kind—proved to be unproductive. It was not so much that the principles of partnership, strength, and negotiation that he laid down were neglected, as that the kind of explicit linkage to military posture and structure that might have been expected to develop simply did not emerge. One key cause seems to have been a combination of high-level opposition in the Pentagon to the outside review of military programs in detail, and a failure at NSC level to develop techniques that could exert higher guidance on such programs without usurping essential internal management functions in the Pentagon.

Much, therefore, still remains to be done to assure that military power is in fact the instrument or agent of policy—and not its determinant, or an independent force in itself. Military power must be understood and prescribed for in a complex higher political context of values, goals, priorities, and constraints. At the same time, we cannot remind ourselves too often that our military forces are not without specific internal imperatives of their own that must form an input to policy if it is to be sensible and realistic—simple operational imperatives of weapons and tactics, for example, such as speed, range, and concealment; or feasibility of logistical support and training; or provisions for command and control. All these impose requirements of readiness, foreign as well as domestic basing, secrecy in plans and intelligence activities, steadiness and practicality of programs, and existence of well-established channels and processes of decision.

The balancing of such external and internal factors—higher security purposes or constraints with internal practicalities and operating requirements—is therefore inescapable. It lies at the heart of the art and the science of security planning, decision, and, accordingly, the shaping of military role and structure.

Within the defense establishment itself, two features of organization and procedure add to the difficulty of the task of strengthening the determination of military role and posture from the top down, rather than the bottom up. For the massive, complex military forces that we maintain, the principal planning task must be carried out within the major military services—the Army, Navy, Air Force, and Marines—which are responsible for building, training, equipping, and maintaining such forces, and for furnishing them to the major operating commands: the Strategic Air Command, the North American Air Defense Command, the European Command (and NATO), the Pacific Command, the Readiness Command, and other smaller commands, that will employ them. There are two linkages that are crucially important and inherently difficult.

The first is the linkage between the military and civilian echelons. The military structure uniquely is the place where deep professional military expertise is organized and brought to bear across the whole range of military responsibilities—from building and supporting the forces to preparing for their employment and directing them in combat if so ordered. The military officers in these echelons, while informed of higher policy, and responsive to it—and while contributing to its formulation by providing their advice, evaluations, and recommendations, do not establish such policy; they could not do so under our system, grounded as it is on the principle of civilian control. It is the civilian echelon that sets the policy, normally with military aid and advice, but this civilian structure, on the other hand, does not, of course, itself possess military expertise—the vast, intricately interwoven body of military knowledge, experience, procedures, and organization that is indispensable to effective and economical military results, attuned to the values and ideals of our society.

The key task then—which impacts directly on the issues dealt with in this book—is to operate the linkage that ties this military expertise effectively to the policy-determining, decisionmaking higher civilian echelons of government. If the linkage is to be effective, the civilian echelons must provide decisions and policies that have sufficient "bite" in them to shape and guide the size, composition, role, and posture of military forces (together with their supporting budgetary, personnel, and material programs). For this purpose, the civilian echelons must receive—and sensibly consider—the proposals, judgments, and factual information that reflect competent, qualified military grasp of a range of problems that extends from advice on broad security policies adequate to safeguard our national interests, to armed forces appropriate for the support of such policies, and programs needed to provide and sustain such forces—but the governing decisions must be civilian.

The problem is compounded by a second linkage: higher authority must look to the Joint Chiefs of Staff (as a corporate body) for advice on operational matters and to the separate service chiefs—representing their respective services—on matters pertaining to the building, training, modernizing, and support of the military forces. Because of the individual chiefs' direct involvement in the budget process (and in the administration of funds), there is a strong tendency for decisionmaking on force structure—with resulting influence on force role and posture—to gravitate into the services. What this means is that the policy factor associated with the civilian echelon, and the operational factor associated with the joint unified military structure tend inherently for this reason to be underweighted, while internal interests—many quite valid but some often open to question as self-serving—tend to receive priority. No easy remedy is available. The leadership techniques of various Secretaries of Defense have had varying success in dealing with this problem. It may be that a systematic analysis along the lines of the present study—linking force decisions as explicitly as possible to interests, objectives, and policies that are subject to civilian determination (i.e., by the government and the public)—will provide a further measure of assistance. Recognition of the overall problem, as reflected in the 1976 Congressional resolution previously cited, is in itself an important step toward its solution.

But if an approach along these lines is indeed adopted and pursued, even if only in principle, what are the chances that evaluations and conclusions along the lines provided in the previous chapters will in fact be found acceptable and valid by responsible governmental authorities? Here two principal comments need to be made. The first concerns the analysis of the external factors involved—the world environment, our national interests and security policies, and the logical connections between these and the specific issues of force role and structure. The second concerns the combining and balancing of these external considerations with internal imperatives such as those that were cited in Chapter 5.

Regarding the latter point, the internal needs and drives of the military services—for continuity of established programs, for undiminished budgets and strength levels, and for innovation and technological advances—exert strong pressures that can readily come into conflict with policy-level determinations. Long years and massive expenditures may be needed to produce a military capability—such as the U.S. antiballistic missile developed in the 1960s and early 1970s—which will then be abandoned because of new developments. Such might include a new technology available to our principal adversary such as multiple warheads that could saturate the defenses or a change in the policy evaluation of the weapon's destablizing effects, e.g., an impact on the force balance that could set off a competitive missile-expansion spiral, or an impact on the readiness and reaction postures of the two sides, which could put a premium on dangerous hair-trigger alert and response. Thus while internal service needs and objectives are important, they cannot be allowed necessarily to govern. The famous encounter in the early 1950s between Secretary of Defense Charles Wilson and the U.S. Congress made the record clear that "what is good for General Motors" was not to be regarded as necessarily and incontestably good for the country.

Such differences are by no means unique to the military services (although they as the builders, trainers, and suppliers of the forces are most directly involved). Also at Department of Defense level, in other Executive Branch departments and agencies, in the White House itself, and certainly in the Congress, the narrow, more self-focused interests of individuals, bureaucratic partisans, political parties, and other constituencies come strongly into play. A whole recent field of defense literature—centering on the behavioral influences and determinants affecting policy and program decisions—highlights the pervasive, sometimes predominant role that has been played by such considerations—in the cancellation of the SKYBOLT missile, for example, or in the decision to adopt the TFX fighter, and in the fight for the B-70. Rarely indeed will such considerations be lacking from the setting of important policies affecting this $100-billion-a-year enterprise. What has been argued here is that they need to be more fully evaluated and counterbalanced—to the benefit of our overall security as well as public and governmental understanding and confidence—by analysis rooted in the broader public interest, as distinguished from more specialized motivations. Governmental organizations inevitably inject their own purposes into program of decisions, even when no actual conflict of interest occurs, and it would be an illusion to suppose that such tendencies can be somehow swept away. Rather, the search must center on identifying what particular internal needs and constraints are truly essential—since these can then be used to condition and temper the purpose-oriented analysis. Distortions that serve solely "parochial interests"—in the term President Eisenhower sometimes had occasion to use—can then better be identified and rejected. Both approaches—the one based on higher purpose or mission, and the other (the so-called behavioral) based on internal imperatives and preferences—have essential contributions to make. So long as the latter (the internal) are taken sufficiently into account to keep programs realistic and efficient, it would appear that the former (the higher purposes) should be dominant.

We come then to the other point raised at the beginning of the chapter—if such an approach is in fact pursued, what is the likelihood that evaluations and conclusions such as those developed in the present study will be found acceptable and valid by the responsible governmental authorities? Here the important question is not of course the particular judgments of this particular author; these are obviously seen through only a single set of eyes, and—although prepared with every effort at care and balance—necessarily reflect personal views. The more fundamental question is how well the U.S. government is prepared, at its current state of structure and effectiveness, to carry out such determinations. Here the answer cannot be wholly encouraging.

The three-year study effort completed in 1975 by the Commission on the Organization of the Government for the Conduct of Foreign Policy (the "Murphy Commission") identified serious and numerous shortcomings in the capacities of the Executive Branch and of the Congress to deal with these matters and to work constructively and effectively with each other. Despite the submission by the Commission of a lengthy list of well-considered recommendations, reflecting the results of a massive supporting study effort as well as nationwide hearings and public discussions, essentially nothing was done within the government following the submission of the report. In part, the inaction may be traceable to reasons such as the post-Watergate, post-Vietnam hiatus in government, the preelection let-down during 1976, and the reluctance and inability of the government (Congress and Executive alike) in these circumstances to face up to difficult, complex issues of foreign and security policy such as SALT II, energy policy and programs, and policy for the third world (as in Angola) or to break away from a concentration on imminent or dramatic crisis negotiations to pursue longer-term objectives. A major product of the Murphy Commission work was its emphasis on the need for more far-reaching, deeply based attacks on major problems of security and foreign policy, and for more effective operation, control, and direction of the major agencies of government concerned with them: the State Department, the Defense Department, and the CIA, in particular. There was little response to that emphasis.

Undoubtedly this remains one of the major security challenges before the government and the nation, and its importance is enhanced—or aggravated—by the prospects and possibilities of future change and challenge that will test and modify the assumptions and premises on which the major lines of U.S. security policy and program effort (including those proposed in this book) are grounded. This is the final set of questions we shall briefly consider.

At the beginning of this chapter, our second question inquired as to the likelihood that international events and circumstances will develop along lines as described or foreseen in these pages. Our concern in this connection runs first to events that are high in their potential for harm if they were to occur (even if their likelihood is low), but also to foreseeable changes in the nature and priority of our security problems in the years ahead, and to shifts in the relationships of major world nations that would endanger and upset world balance, particularly the military balance between major world power centers—the USSR, Europe, the United States, and China. The question covers both the kinds of change that grow and develop over a substantial period of time (even though their effects or their perception may burst on us rather suddenly) and those abrupt reversals, outbreaks, or overturnings that transform the international scene and balance almost overnight.

The changing content of our security concern results primarily from evolving problems of the first type—those that grow and develop over a substantial period of time. Potential challenges to American security via economic channels—vulnerability to oil embargo and dependence on minerals from foreign sources such as bauxite and chrome are prime examples—are threats not amenable to influence through the traditional uses of military force or gunboat diplomacy. The prospects for rising tension along a North-South world axis over terms of trade and economic assistance affecting a coalition of impoverished, third-world nations with burgeoning populations pose a new kind of security threat, likewise ill-suited to military measures available to the United States, even though such might be made unavoidable by third-world resort to violence or disruption of peaceful trading practices. The worrying prospects that we see give even further emphasis to the need to develop an up-to-date, dynamic set of policies toward the third world. The policies needed must guide the nonmilitary efforts and initiatives toward constructive relationship and harmonious interaction (which are the processes that are very much to be preferred), as well as preparation for the fall-back military option, if it should ever need to be exercised, despite all our preferences to the contrary.

Yet along with these new threats to security and peaceful development, the old military threats—some of them a continuing source of awesome danger, such as the nuclear confrontation, some of them a continuing requirement of vigilance and collective effort, as in NATO—retain their former importance undiminished. And the domestic implications and interests bearing on them—illustrated, for example, in the congressional action denying funds to Vietnam, to Angola, and to Turkey, in response to attitudes and internal constituencies in the United States—seem to be reaching deeper into our society, and evoking stronger internal pressures, representations, and constraints. The impairment in national leadership and public confidence in government experienced in recent years in the United States has diminished our ability to place such events in a perspective that shows their bearing on longer-term U.S. security and economic interests and policies. The requirement to do so is clearly one of the consequences of this broadening scope of security issues.

On the world scene, three possible future developments, of varying probability but of vast potential for harm, merit specific mention. The first is the steady apparent slide toward nuclear proliferation, which threatens the introduction of major and systemic instability into the world security environment. The long and growing list of those who have the nuclear weapon or its equivalent (the United States, the USSR, the UK, France, China, and India) plus those who could produce it soon, or stop just short of final assembly (among others, Sweden, Israel, Pakistan, Iran, several Arab states, several Latin American states, and South Korea and Taiwan if they were to choose to do so) is enough to show the wide reach of potential effects. And the problem is by no means limited to the development of weapons by governments. Beyond this is the possibility of seizure of weapons by clandestine means, and particularly by dissident or violent groups. The dangers that the weapons will be put to use tend to rise geometrically as the numbers of nuclear-armed adversary relationships around the world multiply.

One early consequence of such proliferation seems almost certain to be an attempt at withdrawal on the part of the superpowers and their principal allies from too close a connection with those potentially involved, and their adoption of more unilateral and even isolationist lines of policy. Nevertheless, the strong U.S. emotional and economic ties that exist in many cases (with Israel and Iran, for example) forewarn of increased future complications and more dangerous nuclear involvements. So long as Japan and the NATO nations (other than Great Britain and France) show no sign of reaching for the bomb, there is no evident reason why the basic structure of major security arrangements to which the United States is party cannot continue. But should Japan or West Germany decide to "go nuclear," wholly new arrangements would have to be foreseen, in which it appears extremely unlikely that U.S. forces would continue to be maintained in Europe or the Japan-South Korea area. The reason is simple: there would otherwise be unacceptable risk that a *de facto* trigger on the U.S. nuclear arsenal (and thereby on the Soviet as well) would have been created. In this regard, the withdrawal of U.S. (and NATO) forces and installations from France, which actually took place upon the demand of de Gaulle in 1966, would probably in any event have been an inevitable consequence of the French development of an independent nuclear capability not tied to NATO. It is similarly probable that only the special U.S.-UK understandings and the commitment of the British nuclear forces and weapons to NATO obviated the necessity to make comparable withdrawals from Great Britain when the British developed the bomb.

While national commitments to international nuclear safeguards and controls give some degree of assurance that nongovernmental groups will not be able to build bombs contrary to their host government's desires, it does not appear that governments, even of the smaller and less developed countries, could be dependably prevented from doing so once they had determined to go the nuclear weapons route. The more dangerous world of proliferated nuclear capabilities, if it comes into being—and the chances are high that it will—will surely be a major milestone on the road to a quite different and increasingly perilous future security situation.

The tremendous accumulations of capital and currency by lightly populated Arab oil-producing countries are reshaping the power relations of the world. They are providing a time bomb that could detonate with drastic consequences for world security and well-being at some future time. Power is draining away from all but the most dynamic industrial democracies (the United States, Japan, Germany and, to a degree, France); and the weaker economies, especially in the less developed, poorer regions of the world, are being driven into desperate straits that seem bound to place increasing pressure, backed by the threat of potential hostility and outright violence, on the more advanced Western societies. If the oil-rich Arab states remain stable, the possibility of fast-enough adjustment to the new realities seems plausible. But should a Kaddafy seize power in Saudi Arabia, Kuwait, Iran, or the oil-producing emirates, then a new chapter in Western security will immediately be opened. And should Soviet involvement occur, the stakes would be quickly raised to a level at which the possibility of war of worldwide dimensions could not be excluded. But even without Soviet involvement, the possibility of economic attacks and reprisals and of military conflagration would be very high. Such a struggle could easily include, for example, selective impoundment by the Western states adversely affected of foreign funds and investments within their sovereign territories—actions with untold impact on the world trading and financial system on which economic life today is utterly dependent. Again, a profoundly altered set of world security and economic arrangements would be necessitated, and a quick transition into much more militarized conditions and processes of international life could not be excluded.

A third such possibility with profound potential impact on the world security environment would be a rapprochement, even if temporary, between the USSR and China, particularly if it were joined with an agreement between them on spheres of interest, which gave a green light to each to pursue expansion and domination on a continental scale—the USSR to its West and South, for example, and China (and its surrogates) in Asia, the Western Pacific, and the Southern Hemisphere. Although such an arrangement could hardly prove stable—the deep-lying issues and antagonisms between Russia and China as we know them in the present period being in all likelihood far too great to permit this—the possibilities for short-term harm—the subjection of countries now free and the increased risk of the eruption of world war—would be very great. If Soviet-Chinese action along these lines were timed to take advantage of Western disarray or defeatism—following the model, for example, of the Molotov-Ribbentrop pact between Stalinist Russia and Hitlerite Germany in 1939—the prospects could appear even more inviting to the two communist rivals. Such a pact—even without the green light for high-risk diplomacy and resort to military force—would usher in a period of extreme danger for Western peace and security and necessitate a wholly new assessment and projection of Western policy.

These three possibilities obviously do not exhaust all potential eventualities. They are, of course, extreme points on the chart, the full impact of which may never be reached. Short of them, the principal dangers to world security and a stable world order are probably not to be found in foreign lands at all, but in the societies that form the forefront of modern, Western-style civilization. A failure of political systems—and particularly a failure of the self-governing people of the West to govern themselves and direct their countries effectively—would offer the greatest opening for a breakdown of peace, security, and mutual well-being.

But since the rise of modern democratic government, such societies—including our own—whenever they have been faced with grave problems and have been permitted the time to deal with them, have shown a considerable capacity for doing so with a sense of restraint and responsibility. The two murderous world wars of 1914-1918 and 1939-1945 are the principal exceptions in this century. With careful provision for security interests, policies, and forces as we can best now perceive and design them, and with readiness to respond quickly and flexibly to new problems as they begin to emerge, we may have as high a measure of confidence as human frailty will permit that we can continue to meet the new challenges sure to confront us as we march to the future.

Index

Index

Chile, 67, 136

China, 6, 12, 25, 96, 159; attitudes in, 45; communism of, 44, 140, 143; Cultural Revolution, 44-45; frontier of, 41, 56; initiatives by, 139; as nuclear power, 157; and Soviet Union, 27-28, 45, 47-48; and U.S. relations, 32, 44, 156

Chou En-lai, 12

Chromium, 21, 34, 157

Civil defense programs, Soviet, 84

Civil Reserve Air Fleet (CRAF), 145

Civil strife, 122

Civilian control principle in U.S., 153-154

Class struggles, 40

Coal, 32

"Cod Wars," 4

Coercion and threat tactics, 1, 66

Cold War, 50

Collective preparedness and security, 30, 45-46, 55, 75, 87, 99, 107, 111, 138

Colonies and colonial system, 8, 22-24, 137

Commerce and commercial development, 1, 15, 22, 44

Commitments, military, 5, 17, 99

Communications: and command facilities, 61, 78; digital techniques of, 112; integrated networks, 36, 107, 111; sea lines of, 10, 132-135; Soviet, 12; systems of, 95, 101, 104, 128, 141, 144, 146

Communism: Chinese version of, 5, 44 140, 143; international movement, 27, 42; Soviet supported, 3, 13, 40, 50; takeover tactics of, 6, 33, 55, 124, 137; in Western governments, 46

Computer analog techniques, 26, 112

Conakry, 23, 26, 34

Conference on Security and Cooperation in Europe, 70

Confidence, base of, 28, 92

Confrontations: East-West, 61; militant, 72, 85; nuclear, 66; racial, 34-35; religious, 24

Congo, the, 8, 23

Congress, U.S., actions of, 14, 47, 76, 83, 98, 118, 148-155

Conscription, 46, 108

Conventional arms and forces, use of, 55-61, 64, 75, 87, 104, 108-109

Cooperation, patterns of, 39, 137

Copper, 34, 36, 137

Counterforce capabilities, 57, 61, 86, 88-89

Counterinsurgencies, 137

Crises: management, 56; negotiations, 26, 77, 156; short-term, 11; situations, 61, 83-85, 122, 131-132

Cruise missiles, 8, 83, 93-97. *See also* various types of missiles

CSCE, 29

Cuba, 23, 42, 44, 139; in Angola, 48-49, 64; military buildup, 27; and Soviet Union, 35

Cuban missile crisis, 7, 18, 28, 53, 61, 64, 67

Cultural: exchange, 26, 53; heritage, 1, 141; revolution, 44-45

Culture, native, 18, 22, 68, 138

Cyprus, dispute over, 47, 106, 117, 122

Czechoslovakia and the Czechs, 18, 26, 30, 42, 51, 59, 62-63, 100

"Damage-limiting" objectives, 88-89

de Gaulle, General, 114, 158

Decision making processes, 2, 6, 11, 62, 94, 125, 153-154

Defeatism, attitude of, 16

Defense: costs of, 61; forward, 104, 106-107; NATO policy goals of, 29, 75-76, 105; objectives of, 53-59, 102, 112, 114; requirements of, 139; Soviet program for, 73

Defense, Department of, 82, 88, 118, 126, 156; Secretary of, 116, 141, 154-155

Defense Authorization Act, 150

Delivery systems, 127

Delta class submarine, 82

Democracy, Western Style, 5, 12-13, 45, 137-138, 158

About the Author

Andrew J. Goodpaster, a 1939 West Point graduate, served in combat in Italy in World War II and in strategic planning and policy posts in Washington following the war. After receiving the Ph.D. from Princeton University in 1950, he served at SHAPE in Paris and then for six and a half years as a staff assistant on security matters to President Eisenhower in the White House. After commanding an Army division in Germany, and serving again in Washington as assistant to the chairman of the Joint Chiefs of Staff, as director of the Joint Staff (with added duty as U.S. Army Member of the UN Military Staff Committee), and as commandant of the National War College, he concluded his active service by serving for a year as Deputy U.S. Commander in Vietnam and for five and a half years as NATO's Supreme Allied Commander in Europe. Upon retirement, he spent a year and a half as a Senior Fellow at the Woodrow Wilson International Center for Scholars in Washington, and a year as John C. West Professor of Government and International Studies at The Citadel in Charleston, S.C. In June 1977 he was recalled to active duty to serve as superintendent of the U.S. Military Academy at West Point.